DAVID FLUSFEDER has pub... The Gift and, most recently, Jo... of Lovers, written in collabora... Springer, received its UK pre... television critic for *The Times* and a poker columnist for the *Sunday Telegraph*. He teaches Creative Writing at the University of Kent.

Praise for *Luck*:

'In *Luck*, [Flusfeder] bypasses the scientific harsh truth about randomness and probability and instead has written a book about the human side of luck ... Eccentric, insightful meditations' *New Scientist*

'Fascinating ... An eminently enjoyable and engrossing page-turner' *The Jewish Chronicle*

'Flusfeder stands up for superstition' *Telegraph*

'Ruminative ... page-turning' *Times Literary Supplement*

'*Luck* is a wonderful intermingling of the historic, philosophical and literary, with tales of the author and his engaging, complicated, extraordinary father. A joy'
PHILIPPE SANDS, author of *The Last Colony*

'An extraordinary collection of insights into luck, skilfully combining personal stories and historical studies into a partly random structure. It has a glorious unpredictability, producing a stimulating feeling of uncertainty about what the next gem will turn out to be'
DAVID SPIEGELHALTER, author of *The Art of Statistics*

'Thrilling, intelligent and wilfully unique, with the bonus ball of being unexpectedly moving, David Flusfeder's thirteen investigations are the result of a lifetime of original thinking. I loved it'   JAMES RUNCIE, author of *The Great Passion*

'This deep, particular and witty examination of the nature of luck and its role in human existence is an absolute joy, from random start to mysterious finish. A most unusual book'
    LOUISA YOUNG, author of *Twelve Months and a Day*

ALSO BY THE AUTHOR

John the Pupil
A Film by Spencer Ludwig
The Pagan House
The Gift
Morocco
Like Plastic
Man Kills Woman

Luck

A Personal Account of
Fortune, Chance and Risk
in Thirteen Investigations

DAVID FLUSFEDER

4th ESTATE · *London*

4th Estate
An imprint of HarperCollins*Publishers*
1 London Bridge Street
London SE1 9GF

www.4thEstate.co.uk

HarperCollins*Publishers*
Macken House, 39/40 Mayor Street Upper
Dublin 1, D01 C9W8, Ireland

First published in Great Britain in 2022 by 4th Estate
This 4th Estate paperback edition published in 2023

1

ISBN 978-0-00-824527-6

Set in Sabon LT Std
Printed and bound in the UK using 100%
renewable electricity at CPI Group (UK) Ltd

For Felix, Deborah, Michael and David

Though I have always diligently sought for the truth, yet I fear that the recesses in which it is hidden, or my own preoccupations, or a certain dullness of mind may have sometimes stood in my way, so that often in my search for the thing I may have been bewildered by false lights. Therefore I have treated these matters not in the spirit of one who lays down the law but as a student and investigator.

Petrarch, *The Life of Solitude*

Contents

Introduction   1

1   The Philosopher   9
2   Origins   32
3   The Wit of Thomas Bastard   37
4   Roulettenburg   58
5   Getting God to Speak   93
6   My Lucky Underpants   117
7   I Saw Dangeau Play!   140
8   The Slopes of Vesuvius   160
9   The Measurement of Uncertain Things   184
10   By Paths Coincident (canned chance)   208
11   The Rice of Chance   236
12   Group Luck   259
14   Friday the Thirteenth in Las Vegas   277

Acknowledgements   296

Introduction

Any book could have been many other books. This one is my first attempt at non-fiction and it began, simply enough, with a meeting with an editor who was setting up a new imprint that would publish 'literary' non-fiction. Would I, she asked, be interested in writing a book about poker? Of course I was interested. I used to play a lot of poker, mostly tournament poker, and occasionally at a high level. For two years I wrote a weekly poker column in a Sunday newspaper, the material for which I gathered by playing poker in cardrooms legal and semi-legal, and sometimes illegal. I like writing about poker: it lends itself to description, the way the world shrinks to the dimensions of a table, where if the player in seat one is replaced by a new player, one struggles to remember who their predecessor had been, because poker reality is a perpetual conditional now (*If I should, would she...? If he did, what would I...?*), where the past has little place. Poker is a game of incomplete information and every poker hand is an exercise, like a thriller, in the release of further information. And then the pot is awarded, the dealer riffles the deck, and the previous event is over, because poker, like luck, has no memory.

Reluctantly I decided that the world did not need another literary poker book. There are some very good ones already, written by Al Alvarez and James McManus and Anthony Holden. Alvarez's *The Biggest Game in Town* was published

1

in 1983 and the poker world has gone through many generations since then (every four or five years, there's an immense upheaval in theory and practice), but it's something that can only be added to, not supplanted or improved upon. In his later years, whenever he published a new book, Alvarez would have as his apologetic, weary epitaph the same quote from Ecclesiastes 12:12, *Of making many books there is no end, and much study is an affliction of the flesh*. If that is so, and it probably is, let's not make books that have no need to be.

Nonetheless, one of the alternative texts contained within this one is that ghostly book that Ros offered to commission. Poker is often used as a metaphor for life. It can sometimes feel as if life is a metaphor for poker. At the very least, poker, with its precision of analysis and language, can be a very good tool for understanding the world, its ethics, its procedures, its rules to live by. *Decisions, not results* is a central poker wisdom: this hand is one of a near-infinite series; just because your opponent happened to outdraw you and you lost all your money doesn't mean that you acted incorrectly. Poker is anti-teleological. Money in poker is the way to keep score, but if you want to think clearly about how to play, then you have to bracket off the outcome.

But, I thought, and continue to think, albeit in an increasingly despairing, proliferating way, that what the world could do with is a book about luck. It is a subject that every poker player has a relationship to, that every *person* has a relationship to. So, the book that was commissioned was to be a narrative non-fiction book about luck. It would have an 'I', who approximated to its author, who would embark on a journey that might stand for the reader's own, in which he (I) would use his own history to examine, with clear-eyed pitilessness, his relationship to luck, and along the way (scrapes!

adventures!) would uncover some important truths, not least how to maximise the beneficial effects of luck and minimise its malice. *What*, it would ask, *is luck? Is there actually any such thing?* And, *how do I become luckier?*

That book is here also, and it was the one I might have tried the hardest to write. Is there a force that intervenes to hasten or thwart desire and its consummation? How does luck, or 'luck', relate to risk, to chance, to opportunity, to randomness, to destiny, to fortune, to fate? Are these all merely synonyms? Why do some languages – German, Latin, the Scandinavian ones – have no word that separates 'luck' from 'happiness'? Where does God come into this? What about maths? Is it all about numbers, and odds, and big data? And if so, how does that account for some people seeming to be indisputably luckier or unluckier than others?

I normally avoid the presumptive 'we' – the sort of universalising, essentialising usage that magnifies an aspect of the writer's own wavering self, that asserts a generality in place of the particular, that makes a moment into a trend, allows an individual to pretend to represent the crowd. But my belief in my own relationship to luck – or rather my set of questions about my own relationship to luck – is so strong that I refuse to accept that yours can be any weaker. To begin to understand my place in the world, I have to understand my relationship to luck, and before I can do that, I have to look at my parents, my mother's superstitions that so strictly determined how she was able to live, my father's circumstances between the ages of seventeen and twenty-two, when the odds against his survival were so steep as to make it statistically negligible. This is in some respects a family story, how we become, or at least separate ourselves from, our parents.

If it was going to be that, I would have to describe my grandmother's flat in Clapton, something of my father's youth in Warsaw, and later, when I was in my early twenties and he was in his early sixties, the boat trip we took to Atlantic City where he introduced me to my first casino. That book, these books, proliferated unhappily, the wreckages started to pile up, I began to feel like Casaubon, the inadequate scholar in George Eliot's *Middlemarch*, writing, and never finishing, the *Key to all Mythologies*, and on whom the passionately intellectual Dorothea throws herself away. Most religions and cultures have a foundation or origin myth that relies on the process of chance. Does that mean that in order to attend scrupulously to the subject, I had to look at every origin myth as well as my own?

Certainly I would have to study probability theory, and the Epicureans and the Enlightenment, and the history of the lottery, and Victorian literature, and augury and superstition, and philosophy, ancient and modern. I would have to investigate gambling of kinds other than poker, including roulette and the stock market, and before I went Casaubon-mad and dried up entirely, all my Dorotheas wilting beside me, I would have to find some working constraints and procedures.

Notions of luck are so nebulous, the topic slips into so many areas without its lineaments ever being quite satisfactorily drawable or definable, that my own procedures were very hard to find. In the course of pursuing this book, many people helped me in my thoughts on the subject. A fund manager who had patiently tried to explain how financial markets and spread betting worked asked me, 'How's your book on risk getting on?' An artist said, 'How's the chance book?' 'Luck,' I had to remind them. 'It's a book about luck.'

Not only does this suggest that people remember what they are interested in remembering, I also took it to mean that no one really understands what luck is, or agrees exactly on what the word denotes. There are so many constituent parts to the idea of it, which I was going to have to tease out and identify.

As well as the broad, Casaubon questions, there were narrower ones, which seemed fruitful as ways of exploring everything around the subject by focusing on a single point, such as how the novelist Dostoevsky finally managed to quit gambling one night in Germany in 1871, or why the philosopher Wittgenstein took to signing off much of his correspondence with a variation on 'Good luck' in the spring of 1939. By answering these questions, maybe I could begin to find answers to the larger ones.

The alternative books heaped around me as I took airplane rides, as I sat in libraries, as cardboard packages from antiquarian internet booksellers piled up at home, as I wrote up my investigations into hundreds of pages of notes, as I made bets online and off, as I battled with superstition, my own. You can see some of the wrecked books here, the history of an idea, notions of chance in the Western traditions, the gambler's journey, visiting sites such as Baden-Baden where Dostoevsky lost at roulette and abased himself again, as I looked for a way of telling the stories of other people's victories over luck and defeats by it.

*

I had intended to introduce the themes and scope and ambitions with the exemplary tale of Thomas Bastard. Bastard (*c.* 1566–1618) was a poet and epigrammatist, who had been raised from relatively low Dorset origins into becoming a Fellow of New College, Oxford, a society darling, the coming boy, who had issued a poetic defiance to Fortune ('Fortune, I shal braule with thee…'!), whereupon all things went wrong for poor Bastard: suspected of being the author of a set of verses that lampooned the peccadilloes of the Oxford establishment, he was cast out of the university, returned home, where he failed to make a living as a curate and died, destitute, in a Dorchester madhouse.

And then I changed my mind; this was too melodramatic an opening. (And, I discovered afterwards, pursuing further researches into Thomas Bastard, that he had made his challenge to Fortune *after* everything had started going wrong for him.) Instead I would begin with an account of the two poker games that the physicists Niels Bohr and Werner Heisenberg played on a skiing holiday in January 1933. This would be an elegant way of introducing poker itself, a game of skill which uses the mechanisms of chance for its procedures and which was going to recur throughout the book, as being its author's primary way of engaging with risk, luck and hope. And the southern slope of the Großer Traithen in the Eastern Alps is pleasantly scenic as well as emblematic, a scene – physicists, atomic energy, bluffing – which looks to the future, while also saying, or at least suggesting, a lot about language, confidence, courage and luck.

Rethinking again, I thought rather that I should begin with background and back story – that could be the best way into the material, both for reader and writer, the place ascribed to Chance and Fortune and Luck in the mythological foundations of our world and, therefore, all subsequent philosophy and science and art; a line could be drawn from Greek and Roman and Chinese and Hindu and Norse thought all the way to where we are now, narrowing definitions, an introduction as a kind of historical survey.

If my researches have taught me anything, it is that history doesn't work like that: there are no straight lines between then and now and what is to come. I worried also that the chronological approach would be too dry, I did not want to deter readers from this book, or make it dull for me to write – books should not plod, and certainly luck never does, neither good nor bad; if we were to find a verb to account for its movement, it would most accurately be *to swerve*. I thought then it would be best to begin personally, a first-person account of my own relationship to the concept of luck. I should include only those aspects of history and mythology that have fed into my life and education and upbringing. I could begin with my father's last words to me or go back twenty-five years before that, when he took me to Atlantic City; or earlier than that, the other side of my family, the superstitious rules that my mother and grandmother lived by.

But this is not a memoir. Ultimately I decided that, in a book about luck, the best way to organise the material would be to let the dice decide.

I organised my investigations into subject areas, which I would call chapters. Each of these would contain at least one exemplary figure of luck on some aspect of whose life or work I would hang my researches. I assigned these chapters a number from 1 to 14, which I entered into an online randomiser to determine the order of the book that you are now about to read.

CHAPTER 1

The Philosopher

Mrs Bevan was initially somewhat frightened of
Wittgenstein, especially after their first meeting,
which was something of an ordeal for her. Before
Wittgenstein moved into their house [in 1951, the
year he died], Dr Bevan had invited him for supper
to introduce him to his wife. She had been warned
by her husband that Wittgenstein was not one for
small talk and that she should be careful not to say
anything thoughtless. Playing it safe, she remained
silent throughout most of the evening. But when
Wittgenstein mentioned his visit to Ithaca, she chipped
in cheerfully: 'How lucky for you to go to America!'
She realised at once she had said the wrong thing.
Wittgenstein fixed her with an intent stare: 'What do
you mean, *lucky*?'

Ray Monk, *Ludwig Wittgenstein:
The Duty of Genius*

Before March 1939 the philosopher Ludwig Wittgenstein
would generally sign off his correspondence with the phrase
'Good wishes', 'Yours sincerely', or, on occasion, showing his
fondness for pulpy American thrillers, with a jaunty 'So long'.
His philosophical career was conducted largely in silence,
interrupted by an effortful, sometimes agonised, thinking in

public. In his former student Norman Malcolm's *Memoir*, there is a very good description of a Wittgenstein 'lecture' in 1938, in which thirty students and devotees squeezed into the Canadian student James Taylor's room at Trinity College, Cambridge. The only prerequisite to being in Taylor's room was the commitment to attend all of the year's lectures and, in a few rare cases, that the professor liked the student's face. The audience would be called upon to listen, sometimes to answer questions to advance the argument, or, often, just to watch Wittgenstein thinking. 'There were frequent and prolonged periods of silence, with only an occasional mutter from Wittgenstein, and the stillest attention from the others. During these silences, Wittgenstein was extremely tense and active. His gaze was concentrated; his face was alive; his hands made arresting movements; his expression was stern. One knew that one was in the presence of extreme seriousness, absorption, and force of intellect.'

Wittgenstein's custom was to go immediately afterwards to the cinema, as if to a 'shower bath', where he would sit in the front row, to be released in some way by immersion in the spectacle, munching on a cold pork pie or bun he had bought on the walk over. He enjoyed westerns and musicals, Carmen Miranda, Betty Hutton. 'He liked American films and detested English ones. He was inclined to think there *could not* be a decent English film. This was connected with a great distaste he had for English culture and mental habits in general.'

Wittgenstein had tried living away from Cambridge but he kept coming back to it. He had been at the university before the First World War, and was encouraged to return by Bertrand Russell and John Maynard Keynes, with a draft of his *Tractatus Logico-Philosophicus*, which he had composed from notes written at the front when he was a soldier in the

Austrian army, and which became his only completed book. He had left again, given away his inheritance to try to simplify his life; he had been a schoolteacher, an architect, an aviation engineer, had lived in Norway, and in Ireland, as well as his home city of Vienna; but, made into a German citizen by the annexation of Austria in 1938, he accepted the professorship of philosophy and resumed his academic life in Cambridge.

And, on 15 March 1939, in a letter to his friend, the Marxist economist Piero Sraffa, he signed off with,

I wish you a good luck in every way!
Yours
Ludwig Wittgenstein

This most analytical of philosophers, preoccupied with utterance and its limits of meaning, whose own work was, in an effort of spiritual engineering, intended to be of a precise piece with the way he lived, was referring to the hazy notion of 'luck'. Why? It wouldn't be sloppiness, of thought or expression: as Malcolm writes, 'Primarily what made him an awesome and even terrible person, both as a teacher and in personal relationships, was his ruthless integrity, which did not spare himself or anyone else.'

It wasn't the first time he had referred to luck in his correspondence. In 1919, in a letter to Ludwig von Ficker, who would become the German-language publisher of the *Tractatus* (and who helped him distribute his share of the family fortune to more deserving cases such as the poet Rainer Maria Rilke, the painter Oskar Kokoschka and the architect Adolph Loos), he had written,

The MS that I'm sending you now isn't the actual one intended for printing but one that I've only glanced through: it will serve to orientate you, though. The MS intended for printing has been carefully revised. At the moment it is in England with my friend Russell. I sent it to him from the prison camp. [As an Austrian soldier, Wittgenstein had been interned in a POW camp at Monte Cassino.] But he will send it back to me in the near future. And so I wish myself, for once, the best of luck.

Best wishes from yours sincerely
Ludwig Wittgenstein

In a similar vein, in September 1938, he wished his acolyte Rush Rhees '*lots* of good luck with your writing'. Rhees had been struggling to finish his thesis. Wittgenstein suggested, 'Just stick to it; and, *if possible*, sacrifice coherence sometimes. I mean, if you feel you could just now say something, but it isn't exactly the thing which ought to come in *this* place – rather say it and jump about a bit than stick to the "single track" and not get on. That is, if you *can* do it. If you *can't* jump, plod on.' Unable either to jump or to plod, Rhees did not finish his thesis.

But his March 1939 letter to Sraffa was the first time Wittgenstein had closed a correspondence with a reference to luck; and he would continue, often, to do so. There is nothing in the contents of that letter to indicate why this should be: it contains a quotation from Spengler, whose significance Sraffa had failed quite to grasp. It also wonders, in a typical Wittgensteinian mood of prickly austere neediness, if Sraffa had stopped liking being in contact with him. What has luck to do with any of this? Or with Rhees's difficulties with his thesis? Or with the destiny of his book? Or with the writing of this one? Exactly what, to echo Wittgenstein's question to his last hostess Mrs Bevan, did he *mean* by luck?

*

I had intended the Wittgenstein chapter to come late in this book, perhaps to be the penultimate one. After reaching a set of conclusions, tentative or not, about the nature of luck, I might use it to slyly call the entire procedure to account, suggesting a problem with the questions we ask about luck, the terminologies that we use. And, before the climax of the whole book, it would leave us with an image of the lonely philosopher standing in a fairground with his eyes closed, an image of submission and hope.

The randomiser had decided differently. I've got lucky with the placing of the final chapter, but maybe a little unlucky with this one. The investigation of Wittgenstein's use of the word 'luck' will have to serve also as a kind of introduction. Chapter 3, on Thomas Bastard, will bring in a prehistory of ideas from Luck's classical precursors, Fortuna and Tyche, as well as being a case study of a man who went to war with Fortune. This one will have to introduce some of the concepts and themes of my investigations: private meanings, unpredictable movements, hidden patterns, the importance of the single still moment, and the limits of induction.

Wittgenstein's acolyte Norman Malcolm recollected, 'LW once said that a serious and good philosophical work could be written that would consist entirely of *jokes* (without being facetious). Another time he said that a philosophical treatise might contain nothing but questions (without answers).'

So was the philosopher's sign-off a kind of joke? Or a question without a question mark. Or utterly without meaning? A verbal tic or flourish that he had picked up, maybe during one of his stays in Ireland? One wants to find a trigger for him, an event that produced this movement of thought that had the consequence of a new habit of expression. Might it even have been a private joke with himself, who had once had the childhood nickname 'Lucki'?

In statement 116 of *Philosophical Investigations* he wrote,

> When philosophers use a word – 'knowledge', 'being', 'object', 'I', 'proposition-sentence', 'name' – and try to grasp the *essence* of the thing, one must always ask oneself: is the word ever actually used in this way in the language in which it is at home? –
>
> What *we* do is to bring words back from their metaphysical to their everyday use.

Let's look then at the Wittgensteinian everyday. The day he wrote that letter to Sraffa, Wittgenstein had been lecturing about 'one-to-one correlations'. This was one of the lectures that would later be written up by students and published after his death as *Remarks on the Foundations of Mathematics*. They seem to be staged, with the young mathematician Alan Turing cast as a sort of straight-man antagonist, and others, such as Casimir Lewy and the ironically named John Wisdom, as pitiably simple children, to demonstrate the limitations of mathematical and scientific procedure and to disprove the scientific basis for much of mathematics.

It was in these kinds of situations, in James Taylor's room, establishing the rhythms and limits of what can and cannot be said, despite his frustrations with conveying his meaning, and his contrary turns of mind, that Wittgenstein was at his surest.

In the lecture he gave that day, Wittgenstein denies the infinite as a possible property of sets of numbers. This is consistent with the position of 'the early Wittgenstein', who questioned the 'probability' of, as David Hume had put it, proceeding 'upon the supposition that the future will be conformable to the past'.

The idea of induction, or, rather, a rejection of the procedure of inductive reasoning, is going to recur throughout this book. Just because event B has happened before, doesn't mean that it will happen again. Just because event A preceded event B the last time that B occurred, doesn't mean that B is caused by A. And let's not invent an event C that will inevitably follow on from A and B in an unknowable future.

Things do not happen because we expect them to, or want them to. It might be more appropriate to examine the expecting and the wanting rather than the hoped-for or feared event.

Wittgenstein wrote in the *Tractatus*,

> The so-called law of induction… has no logical foundation but only a psychological one
> It is clear that there are no grounds for believing that the simplest course of events will really happen.
> That the sun will rise tomorrow is a hypothesis; and that means that we do not *know* whether it will rise.

There are only hypotheses, and maybe likelihoods, and things as they are, and the faulty thinking self.

As Wittgenstein had put it in a journal entry,

> When we think of the world's future, we always mean the destination it will reach if it keeps going in the direction we can see it going in now; it does not occur to us that its path is not a straight line but a curve, constantly changing direction.

Out in the world, the curve was making him anxious for his family – in post-Anschluss Austria, the Wittgensteins, despite their baptism, had been classified as Jews and therefore subject to Nazi racial laws. He was anxious for himself, worried about being a refugee – he was still waiting to find out if his application for British citizenship had been granted; and the primary reason he had taken the professorship was to gain, if only temporarily, a safe place to be.

Sraffa, who was also Jewish, was sensitive to the risks his friend would be running should he return home. He had been urging Wittgenstein to take up British nationality and a Cambridge teaching position, and had succeeded in persuading him not to return to Vienna.

So where does luck come into this?

The second time Wittgenstein signed off a letter in that way it again followed an expression of neediness. On 19 May 1939, he wrote to his former student, the New Zealander Raymond Townsend. Here is the needy part of the letter,

> It's nice to know that you'll be thinking of me on the ship – but I want to say something and I mean it seriously: there is such a thing as neglecting your friends. Thinking of them is nice but it's rather easy and doesn't do them much good. They sometimes need advice, sometimes help, sometimes just backing up and if you never see them you give them no opportunity to show you what they need... It may be a nuisance that in this world we have bodies and senses and can't have purely *spiritual* contact with each other; but there's no getting away from this fact. Of course I don't *know* that you haven't *most excellent* reasons for isolating yourself. On the other hand I think it just conceivable that you aren't *quite* aware that you're drifting away from friends (who, as I said, have *bodies*.)...

I wish you lots of good luck.
Yours affectionately
Ludwig Wittgenstein

Was this neediness the key to the Wittgensteinian use of 'luck'? A sort of code to the writer if not also to the recipient?

It seems not: on 29 June 1940, he wrote a brief thank you letter to Norman Malcolm, who had returned to America, and was continuing to send LW the Street & Smith detective magazines the philosopher loved so much. The letter ended with an emphatic,

Good luck!!
Affectionately
Ludwig Wittgenstein

Wittgenstein could never understand why anyone might choose to read mediocre philosophy over good pulp fiction. He used detective stories in a similar way to the films he went to after his lectures, as a form of immersive mental hygiene. He was a particular fan of the comic hard-boiled work of Norbert Davis, especially those stories that featured the crime-fighting team of Doan and Carstairs, a private detective and his Great Dane. He contemplated sending a fan letter to Davis – perhaps he should have: Davis committed suicide, for unknown reasons, in 1940. Wittgenstein's letter may have been the thing that Davis needed, that would have tipped the balance towards life. If so, he was unlucky.

Ludwig received his British nationality; his brother Paul, who had continued his career as a concert pianist even after the amputation of his right arm, found his way to New York; but their sisters were still in Vienna, and Wittgenstein's use of the word 'luck' in his letters increased in frequency. There was a flurry in his correspondence with the physicist W.H. Watson from June to October 1940, 'Good luck!', 'I wish you lots of good luck!!', 'I wish you good luck'.

And to Townsend again, 19 July 1940,

I wish you good luck, outside and inside!
    Affectionately yours,
    Ludwig Wittgenstein

Was he needing something from these men? Frustratingly, if looking for keys is the sort of thinking that thinking should be, the answer is no.

So if it wasn't the content of the letter was it maybe a quality of the recipient that elicited the reference to luck? A sort of private code to denote the young men whose faces he admired?

To G.E. Moore, 7 March 1941,

Please remember me to Mrs Moore. I wish you *lots* of good luck.
    Yours
    Ludwig Wittgenstein

Moore was the second-most highly regarded philosopher in Cambridge at the time, whom Wittgenstein saw as a kind of skilled and admirable but essentially foolish infant. There can be no question of sexual desire here, or of neediness. So is the use of luck here a joke? Or just a flexing of an English idiom that adventurous non-native speakers liked to use in the 1930s and 1940s, along with such words as 'rot' and 'damned' and 'bloody', all of which Wittgenstein had a taste for? For example, 'It is dammed hard to write things that make blank sheets better!'

Piero Sraffa, whose March 1939 letter started this all off, was one of the very few people whom Wittgenstein would ever recognise as an intellectual equal – although Sraffa would eventually, in 1946, refuse to have any more dealings with him: their conversations were too demanding, too tiring and all about Wittgenstein. Another was the precocious good-natured genius Frank Ramsey, who had translated the *Tractatus* into English and had been, notionally at least, Wittgenstein's doctoral supervisor. Wittgenstein wrote in the Preface to the *Tractatus*, 'The whole sense of the book might be summed up in the following words: what can be said at all can be said clearly, and what we cannot talk about we must pass over in silence.'

He repeats the idea, with its echo of dying Hamlet's 'The rest is silence', in the final sentence of the *Tractatus*. Proposition 7 states, 'What we cannot speak about we must pass over in silence.' Ramsey whimsically, and not entirely uncritically, paraphrased this in an early review of the work, 'What we can't say we can't say, and we can't whistle it either.' (Wittgenstein, it should be noted, was an extremely gifted whistler – his architectural colleague in Vienna, Paul Engelmann, wrote, 'On one occasion, when the conversation turned to the viola part in the third movement of a Beethoven string quartet, he whistled the part from beginning to end, with a tone as pure and strong as that of an instrument.')

But Ramsey had also written of him, 'Some of his sentences are intentionally ambiguous, having an ordinary meaning and a more difficult meaning which he also believes.'

As with sentences, so with words. And – to make an un-Rheesian jump here – it is likely that Wittgenstein's use of the word 'luck' had a 'more difficult meaning'.

Wittgenstein had said something similar, in a letter to Ficker,

You won't – I really believe – get too much out of reading it. Because you won't understand it; the content will seem strange to you. In reality, it isn't strange to you, for the point is ethical. I once wanted to give a few words in the foreword which now actually are not in it, which, however, I'll write to you now because they might be a key for you: I wanted to write that my work consists of two parts: of the one which is here, and of everything which I have *not* written. And precisely this second part is the important one.

After Malcolm returned to the US and took up a teaching position at Princeton, Wittgenstein wrote to him,

I wish you good luck; in particular with your work at the university. The temptation for you to cheat yourself will be *overwhelming* (although I don't mean more for you than anyone else in your position). *Only by a miracle* will you be able to do decent work in teaching philosophy.

Is this the answer then? That luck is a kind of charm to protect integrity? The probability of corruption is so great in academic life, maybe even more so than elsewhere, that luck must be invoked to buck the overwhelming odds?

The German word *Glück* means 'happiness' as well as 'luck'. Considering the problem of how to express one's good intentions towards others, Wittgenstein had written in his 1916 journal, 'It seems one can't say more than: Live happily!' When Wittgenstein was wishing Sraffa, and others, good luck, it was a development of that 1916 thought, and contained within it that same friendly spirit of wished-for happiness.

Wittgenstein did not have a gift for happiness. He was disgruntledly living with Francis Skinner, who adored him.

He was trying to finish a book, which he would never finish (if he'd lived longer than sixty-two, he would probably never have finished the *Philosophical Investigations*). His relationship with Skinner was not surviving their living together, no matter how attentive Skinner was, no matter how much in awe and love he was – taking on his idol's habits, this mild former philosophy student becoming ferocious against lesser philosophers' lack of seriousness, and accepting his master's practice of banishing carpets and cleaning the floors by pouring tea leaves on to the wood to soak up the dirt before sweeping it all away.

As Malcolm recollected,

> Undoubtedly the idea of being a *professional* philosopher was very repugnant to him. Universities and academic life he disliked intensely... He believed that his influence as a teacher was largely harmful. He was disgusted and pained by what he observed of the half-understanding of his philosophical ideas, and of a tendency towards a shallow cleverness in his students... He once concluded a year's lectures with this sentence: 'The only seed that I am likely to sow is a certain jargon.'

Malcolm and Wittgenstein had a falling-out because LW was horrified at the younger man's naivety in relation to British military operations, the colonial's trust in the probity of the mother empire. Horrified is not quite the right word; LW's reaction was closer to fury. Malcolm's political ingenuousness demonstrated an awful failure, both on Malcolm's part as a philosopher and on his own as a teacher: In November 1944 Wittgenstein wrote to him, what is the use of studying philosophy if all that it does for you is to enable you to talk with some plausibility about some abstruse questions of logic, etc., & if it does not improve your thinking about the important questions of everyday life, if it does not make you more conscientious than any... journalist in the use of the *dangerous* phrases such people use for their own ends. You see, I know that it's difficult to think *well* about 'certainty', 'probability', 'perception', etc. But it is, if possible, still more difficult to think, or *try* to think, really honestly about your life & other people's lives. And the trouble is that thinking about these things is *not thrilling*, but often downright nasty. And when it's nasty then it's *most* important – Let me stop preaching. What I wanted to say was this: I'd *very* much like to see you again; but if we meet it would be wrong to avoid talking about serious non-philosophical things. Being timid I don't like clashes, & particularly not with people I like. But I'd rather have a clash than mere superficial talk – well, I thought that when you gradually ceased writing to me it was because you felt that if we were to dig down deep enough we wouldn't be able to see eye to eye in very serious matters. *Perhaps I was quite wrong.* But anyway, if we live to see each other again let's not shirk digging. You can't think decently if you don't want to hurt yourself. I know all about it because I'm a shirker... Read this letter in a good spirit! Good luck!

In a notebook of 1938 he had written, 'Whoever is unwilling to descend into himself, because it is too painful, will of course remain superficial in his writing.' And the following year, 'The truth can be spoken only by one who rests in it; not by one who still rests in falsehood, and who reaches out from falsehood to truth just once.' And, from 1942, 'A man will be imprisoned in a room with a door that's unlocked and opens inwards; as long as it does not occur to him to pull rather than push it.'

This is the sort of intellectual and moral passion that excites other people, and attracts acolytes. There is a mysticism here that's close to religion, and which is reminiscent of Kafka's attitude to his own work,

> Altogether, I think we ought to read only books that bite and sting us. If the book we are reading doesn't shake us awake like a blow to the skull, why bother reading it in the first place? So that it can make us happy [*glücklich*] as you put it? Good God, we'd be just as happy if we had no books at all; books that make us happy we could, in a pinch, also write ourselves. What we need are books that hit us like a most painful misfortune [*Unglück*], like the death of someone we loved more than we love ourselves, that make us feel as though we had been banished to the woods, far from any human presence, like a suicide. A book must be the axe for the frozen sea within us.

Wittgenstein was a great reader of Dostoevsky but had no taste for Kafka. He saw him as laboriously creating worlds and characters who are preoccupied with the wrong things. His notion of Kafka was of someone who exerted great effort to ask the wrong questions, whereas much of the force of Wittgenstein's labours went to demonstrating that most questions were wrong.

Proposition 1 of the *Tractatus* is, 'The world is what is the case.' Wittgenstein had been considering Joseph Butler's dictum 'Everything is what it is, and not another thing' as a possible motto for his *Philosophical Investigations*, but, contrary to the spirit of that was his tendency, both in discussion and thought, to argue by analogy: 'What I invent are new similes.'

Here's one, which has the power and mysterious lucidity of a Kafka parable,

> A person caught in a philosophical confusion is like a man in a room who wants to get out but doesn't know how. He tries the window but it is too high. He tries the chimney but it is too narrow. And if he would only *turn around*, he would see that the door has been open all the time!

In a letter to Francis Skinner's friend Rowland Hutt in 1945, when the war in Europe was over and Wittgenstein was despairing about the future, purportedly the world's, but which he had solipsistically connected to his own, he wrote, as nakedly as he ever could, not bothering for once to hide occult meaning behind an apparent one, wishing Hutt 'lots of luck… I really mean: strength to bear whatever comes'.

It comes as a slight disappointment to read this, finally an explicit use of this secret word. The 'strength to bear whatever comes' is a kind of stoicism that again relates to Wittgenstein's notion of decency. Actually, Cynicism is the more accurate philosophical label, the school inspired by Diogenes, the 'dog philosopher'. The story goes that when someone asked Diogenes what he had gained from philosophy, he said, 'This, if nothing else, that I'm prepared for every fortune.'

Just as Diogenes did not enquire into the nature of the workings of fortune, or Tyche, as it would have been called in ancient Greece, merely recommending that the individual be unmoved by vagaries in their state, Wittgenstein was striving, so restlessly, just to be.

In January 1946 he wrote to his sister Helene, who, after lengthy, and expensive, negotiations with the German authorities, had survived the war in Austria,

I'm healthy, as usual – I'm sorry to hear that the house in the Alleegasse has been damaged. But as long as it's habitable, then that, in itself, is good luck indeed.

To Georg von Wright in February 1947, Wittgenstein wrote this rather masterful explanation of why he was going to pay no attention to his protégé's work,

I'm glad that you are going to lecture here, and I know that by attending your lectures I could learn a very great deal. In spite of this I will not come to them – for the *sole* reason that, in order to *live* and to *work*, I have to allow *no* import of foreign goods (i.e., philosophical ones) into my mind. For the same reason I haven't read your book, though I am convinced of it's [sic] excellence. If you think that I'm getting old – you're right. So long! and good luck!

After finally resigning his professorship in 1947, Wittgenstein went to Ireland, which, along with remote parts of Norway, was where he found it particularly congenial to live. He wrote to Malcolm from Wicklow,

It's a little Guest House 2½ to 3 hours by bus from Dublin. It's not *too* bad & I hope I'll acclimatize. Of course, right now I still feel completely strange & uncomfortable. That I haven't worked a stroke for ages goes without saying... I wish you lots of luck & I know you wish me the same. Both of us need it like hell. And other people do too.

And, in July 1948, 'If my philosophical talent comes to an end now it's bad luck, but that's all.'

In 1950, he visited Malcolm in the US. 'My wife once gave him some Swiss cheese and rye bread for lunch, which he greatly liked. Thereafter he would more or less insist on eating bread and cheese at all meals, largely ignoring the various dishes that my wife prepared. Wittgenstein declared that it did not much matter to him *what* he ate, so long as it was always the *same*.'

*

I need more. It is all very well trying to emulate Diogenes and Wittgenstein and remain unbowed in the face of circumstances, but it's not enough just to be resilient and still. I may want to avoid reading my friends' books and attending their lectures, but I don't want to reduce all stimuli and narrow all relations in the interest of remaining undisturbed to eat my invariable cheese sandwich in an uncarpeted room.

I write this sitting at desk number 242 in the British Library, because 42 has become my 'lucky' number. I'm reading Wittgenstein's correspondence in between exchanging Facebook messages with Ashley Revell, who had staked everything he owned on a single spin of a roulette wheel. My usual neighbour to my left is reading a book about the Belsen concentration camp. On my right, a public-school type is flicking between Facebook and Twitter and the sports marketing PowerPoint he is here to make and scanning pages of pastel-coloured polo shirts on the John Lewis website.

Later I will go to the casino to play poker, for the first time in many months. When I got dressed this morning I put on a pair of green underpants that I associate with being lucky at cards. Rationally I know that this is merely sympathetic magic, but all the same, it's not just a definition I'm after with all this. Of course there's a part of me that's the disinterested cultural detective, curious about the notion of luck, excited by the pursuit of it in books and libraries, across cultures – but there's also the card player and occasional gambler who is not just content with adopting a submissive attitude towards luck and chance and fate and variance; I want to be able to influence it, or, if that will be impossible, at least to learn how to maximise its benefits and minimise its ill effects.

And, maybe, the message from Wittgenstein is that I'm going to have to dig deeper into myself than I had thought to do, or wanted to do, expose more than I had intended, to push against the limits of my own thinking and possibilities and courage. I am attracted by the unbending purity of the manner of Wittgenstein's thinking, and I hope to borrow some rigour from him, but it might be, as Wittgenstein wrote, *downright nasty* to have to do this.

*

In January 1950, with the cancer that was going to kill him finally diagnosed, Wittgenstein seemed to be enlarging his definition of luck, 'I am very well indeed now and am anything but depressed. I've had no end of luck. Even that Dr Mooney did not recognise my illness was very lucky.'

By this, he probably meant that he was glad he did not have to endure an operation in the USA and therefore risk dying there rather than in Europe. And perhaps this was also in his mind when he snapped at Mrs Bevan, his final hostess, the wife of the doctor who made the diagnosis of cancer. Wittgenstein's anger at her might not just be at a sloppy use of language. She had said something about luck that was contrary to his own thoughts about mortality and place. And maybe he had a sharp suspicion that Dr Bevan had told his wife more about the eminent patient than he might have?

And, although this could be a stretch, might he be thinking of the other ones who had died before their time, such as the prodigious Ramsey, who suffered from liver problems throughout his short life, and was dead of jaundice at the age of twenty-six after an abdominal operation in 1930; or the Canadian James Taylor, in whose room those Cambridge lectures took place, killed in a barroom brawl on his way to take up a professorship of philosophy at the University of Melbourne in 1946 (a dramatic vindication of Wittgenstein's disapproval of his students becoming academics); or that, before he'd left Vienna, three of his four brothers had committed suicide; and those of his other friends who had died early, such as his likely first English love David Pinsent, killed in a 1918 flying accident at the age of twenty-six; and the luckless Francis Skinner, dead of polio at twenty-nine in 1941.

*

Back in 1939, when Wittgenstein was attempting to dismantle the philosophical method (and when, of course, he was starting to use the word 'luck' to sign off his correspondence), he defined scientific analysis as being a matter of discovering something new. The example he gave was the discovery that water is composed of two Hydrogen molecules to one Oxygen: H_2O – but philosophy doesn't discover anything new; it can't.

I am not a philosopher, except in the sense of Wittgenstein's notebook definition of a philosopher as someone 'who has to cure many intellectual diseases in themselves before they can arrive at the notions of common sense'. I am trying to break the concept of luck down into its constituent parts and so discover how they are arranged and what their relationships are to each other and, therefore, to us. (And by 'us' I of course mean *me*.)

Wittgenstein would have declared that this project is doomed before it starts. All we can do is to look at how the words (hardly even 'concepts' or 'ideas') *luck* and *lucky* and *luckily* are used; we can describe; we may not uncover, we are not going to pull away some obscuring veil to get at an essential truth. As Euripides has Herakles say in *Alcestis*,

> The ways of Tyche are out of our sight.
>   We cannot learn the ways, nor can we make them ours
> by craft.

But I'm going to try.

In happier times, before the war, Malcolm and his wife and Wittgenstein had gone on occasion to the Cambridge fair at Midsummer Common. 'He liked to roll pennies for prizes. He refused to try to direct the course of the penny, even closing his eyes before releasing it, because "everything must be left to chance".'

This is like Engelmann's report of the young Viennese Wittgenstein saying that it was the task of a human being 'to live modestly in the service of others... to give up the arrogant attempt to influence fate'.

And here we might have the nub of it. Luck relates to the external things that the individual has no power to alter, that it would be an indecent arrogance to attempt to influence. But, as with so many things in Wittgenstein's character and interactions with the world, there is a tension.

On his American trip, walking in the New Jersey hills with the philosopher Bouwsma,

> he returned to the way in which we borrow – hints. He had seen a play, a third-rate, poor play, when he was twenty-two. One detail in that play had made a powerful impression upon him. It was a trifle. But here some peasant, ne'er-do-well says in the play: 'Nothing can hurt me.'

That remark went through him and now he remembers it. It started things. You can't tell. The most important things just happen to you.

The most important things just happen to you… the strength to bear whatever comes… And this is what Engelmann wrote about, Wittgenstein enlisting the assistance of luck and chance in the fight against his own arrogance, his effort to become 'decent', standing at the fairground stall on Midsummer Common, with his eyes closed.

His severity was a consequence of his search for perfection. When Wittgenstein and Engelmann were designing Ludwig's sister's house in Vienna, Ludwig had taken a year, of labour and thought and manufacture, to design the door handles and window latches, and another year for the radiators. Just before the house was finished, he held things up, again, by insisting that one ceiling be raised by three centimetres, for the room to be given its ideal proportions. He was always trying to find the right doorknob, the correct way to clean a floor, how to use language, how to think, how to live. Luck, for Lucki – as he humbled himself to bear whatever happened – was an escape from his perpetual moral vigilance.

I wanted to see where Wittgenstein had stood passive before the routes of chance, where he displayed the courage to bear whatever might come as his companions, amused and a little baffled, or maybe used to this, watched him standing wilfully blind at the pennies rolling down their wooden run, or, when it was their turn, coerced by their terrifying friend to follow the same procedure, forbidden from trying to influence the game in any way. I wanted to have a go myself; I too would conquer my instinct to control the coins; I too would stand there, like Wittgenstein, dignified and alone, as the pennies – I imagined old pre-decimal pennies, smoothed and browned by the years, Victoria maybe on one side, Britannia on the other – clattered down the board.

Imagination was necessary because I didn't get to Cambridge in time for the summer fair. I'd made a mistake with the dates, and visited the city a week after the fair was over. I had wanted to come closer to Wittgenstein and had hoped to push away our distance in time (and all our other differences) by connecting somehow in space. I walked on the common, which was still battered by the machinery of the fair, flattened grass, little dips in the earth from the wheels of a heavy fairground ride. Dogs were being exercised. People were enjoying picnics. A young man, who could have been me half a lifetime ago, was reading a book, lying on his back, holding the paperback between himself and the sun. I rested my heel in one of the dips in the earth, and tried to pretend to myself that because Wittgenstein might have stood here too, I was getting closer to an understanding of it all. But he was absent and the fair was gone and I felt a bit silly, I couldn't even call it unlucky, to have missed it.

Cambridge was done with. Thoughts now turned to Thomas Bastard, of Dorset.

I liked what the randomiser had chosen – to go from the fatalism of Wittgenstein, his restless quietism in the face of events, of luck, of himself, to someone who had struck an adversarial attitude to fortune, whose subsequent disasters might be attributable to a single emblematic moment, and whose places I could visit without needing to attend to a timetable. The life story of Thomas Bastard will enable me to trace the history of the idea of luck, to show where it came from. But first, briefly, where I come from.

CHAPTER 2

Origins

There is just a matter of simple coincidences.

Joe Flusfeder

Everything happens for a reason.

Gertrude Tesser

My father was not interested in philosophy. Nor did he favour discussion about abstract categories or actual emotions. What he saw were systems. With great clarity, he would immediately grasp how different elements connected. This could be a complex machine like a record press, or a group of men organised into a factory or a prison camp. He had learned to pursue, or maybe this was an innate capacity that was developed, and rewarded, by experience, his own advantage within the system he happened to be inside.

When I was six, my parents separated, and my mother returned to London, bringing with her my older sister and me. My sister and I became Londoners, but we would spend our summers with my father in New Jersey, and, after he remarried, New York City. My father was the idol and terror of my youth, the smartest and toughest man I've ever known. He was pursuing his new life, a business he was growing, a fresh alliance with a second wife with whom he enjoyed spending money. He was not interested in the inner lives of others, especially not that of a narrow bookish boy. He and I were not, at least until the last years of his decline, 'close'. I was a disappointing mystery to him, and only occasionally would he talk about himself, about his past, those mid-century moments when all the odds were against him. Generally, if it did come out, it would be as throwaway remarks about the anti-Semitism of Poles, the Nazism of Germans, the brutality of Russians, the xenophobia of the English.

Izrael, or Izio, Flusfeder was born in 1922 in Warsaw. My mother Gertrude, or Trudy, Tesser was born in 1923 in London. Both her parents, Lazarus and Sarah, were from the region Galicia, which is now in Ukraine and has been in Poland and Russia. They were from villages, similar shtetls, who had separately made the journey over to London in around 1905 as part of the great westward wave of Jewish dispersal in response to Russian anti-Semitic pogroms. Lazarus Tesser's surname was invented at Tilbury Docks when an immigration officer was unable to spell, or, probably, to hear, a long, multi-syllabled foreign name that was briskly shortened into an approximation of its first two sounds.

My mother grew up in the East End of London, their fourth and youngest surviving child. Lazarus worked, sometimes, as a tailor's presser, and if his name hadn't been mangled at customs, he might have found something else, but Tesser the Presser has such a ring to it that it was probably inescapable. I have one photograph of him, in a cloth cap, one of dozens of men standing on a patch of wasteland, an empty street-corner plot on the Commercial Road. They would have gathered to look for work, but they were talking, kibitzing. Lazarus was engaged with politics – he read socialist tracts, and the *Daily Worker* each morning, and was on the Cable Street barricades against Mosley's blackshirts in 1936. And he liked to gamble. He gambled on horses and he gambled on cards.

Lazarus died before I was born. His widow, my grandmother, spoke a mix of three languages, Polish, Yiddish and English. Unlike her street-corner scholar husband she was illiterate in all three of them. Sarah told fortunes from tea leaves. I was, she declared, an old soul who had been this way before. When my mother first brought me to meet her mother, on a transatlantic trip in the early 1960s, Sarah declared me to be the reincarnation of her husband. Maybe I inherited my liking for cards through that maternal line. I'm happy to believe so. I have always been more resistant to the phrases that my mother and grandmother passed between them as truth. 'Everything happens for a reason', 'Bad things always happen in threes' were two of them. Conversations between Trudy and Sarah often took the form of disputes over the nature of the latest adverse event, whether it truly counted as bad, or if it didn't quite rank as bad enough, which meant that they would have to continue their vigil, to wait for the third disaster to bring the current series to an end. 'Only the good die young' was another, which was gloomily repeated increasingly often by my grandmother as she went into her eighties and nineties.

My parents met at a dance in London in 1946 organised by the Polish-Jewish Ex-Servicemen's Association. My father, in a rare stab at romantic thinking, claimed to believe that he and my mother might have met anyway if things had been different, if not in London after the Second World War, then maybe Warsaw or New York, on any number of alternative historical lines. I don't think he believed this for a moment.

In November 1939, my mother was living, unhappily, in Devon, which an East End education had not prepared her for. She was one of millions of children who had been evacuated from London during the period of the 'Phoney War'. Many homesick evacuees, and those who could work and bring money into the household – my mother satisfied both criteria – returned home in time for the Blitz.

In November 1939, the German army had been occupying Warsaw for nearly two months. My father, then seventeen, was walking home from school. He was the younger of two sons of a middle-class mercantile family. His older brother, David, who I'm named for, was the good son. Izio was the bad son. He was wayward. He had already been expelled from half a dozen schools. He attended meetings of communist cells. His provoked parents paid a well-regarded schoolmate, Benny Zysmanowicz, to try to be a good influence on him.

Entering a square, without, for once, Benny to keep an eye on him, this bad son walked into a round-up. The German occupiers were looking for Jews to fill a work gang. To help them, they had some Polish locals to question the young men they were forcibly enlisting, *Where do you live? What is your father's occupation? What school do you go to?* The Poles told the Germans that my father's answers meant that he was Jewish, and the soldiers directed him towards a nearby line.

Later he would find out why this was being done: to clear swampland outside the city, where there was typhus, and dysentery, to which many of the workers succumbed. The surviving workers would anyway be entered in the system now, which moved towards the genocidal *Aktionen* of 1942. But that was all unthinkable in late 1939. The Final Solution hadn't started, the Warsaw Ghetto hadn't been established. The Jews weren't yet compelled to wear the yellow star. All Izio Flusfeder knew was that a hostile army, whose leaders favoured some very unpleasant propaganda, was occupying his city and that a soldier with a gun had directed him to stand in a line. He tried to take the designated place, but was pushed away. At the head of the line was an older, self-satisfied man, who jerked his thumb behind him. *I was here first*, he said. *Get to the back.*

As my father told the story, he shrugged, walked down the line, and, just as he was about to take his place at its rear, he heard someone speaking to him, *Walk on, you schmuck. Keep walking.* It had not occurred to him to do so, and he did not look around to find the speaker, because he did not want to draw attention to himself, but what was being said was so intuitively true that he knew he had to act on it, so he kept walking, an easy pace, past the rear of the line, and past the rear of all the other lines, and, without ever seeing the face of the man who had saved him, he walked away, out of the square.

When, many years later, he finally on occasion consented to talk about his experiences in the war, all the moments when the odds were so heavily against him, a Jew in Nazi-occupied Warsaw, a forced labourer in the Siberian Gulag, a Polish soldier at the Battle of Monte Cassino, I asked him whether he thought there was some intrinsic quality that had helped him to survive, and he shrugged, maybe as he had shrugged in that moment when he had been told to go elsewhere by the stupidly proud head of the line in the Warsaw square round-up, because he had chosen not to allow the question to have any meaning for him, and he told me that story, to which there was only one moral, *I was lucky.*

It was this story that started my researches into this book, that had me trying to understand the processes at work, and to go looking for exemplary figures of luck and their opposites.

CHAPTER 3

The Wit of Thomas Bastard

Fortune may be mistress of one half of our actions
but... even she leaves the other half, or almost, under
our control.

Niccolò Machiavelli, *The Prince*

Fortune...
I shal braule with thee

Thomas Bastard, 'In Fortunam'

When I was a student I had a period of going down to the
bookmaker's to place bets on horse races. I would smoke
cigarettes and look at the newspaper form sheets pinned to
the wall and unthink myself into a semi-trance to wait for
something to assert itself upon me, the time of a race, the
name of a horse, its colours. In a small sort of way I made
money. This period ended when I went for the first time to
an actual racecourse. Before the meeting began I looked at
the horses in the paddocks with all the certainty of intuition
and made multiple bets on every race, none of which came
in.

It is much easier to make bets these days. In a world where
the majority of leading football clubs are sponsored by online
betting companies, I could take advantage of multiple sign-up

offers and drain the family funds within half an hour (after an itchy wait for identity documents to be verified) without getting out of bed. It occurred to me that by making bets, by risking money along the way of writing this book, I might be able to establish some kind of measuring stick for the action of luck in my life. To begin, I put £10 on Tyche to win the 13.15 race at Wolverhampton.

*

Thomas Bastard never gambled, I'm sure of that, because of a single word that he never used.

He was born, probably in 1566, in Blandford Forum, a market town in north Dorset. His family was from the yeomanry, which was the class of small landowners below the gentry. His aptitude for learning was noticed early, and, with the aid of a benevolent local patron, he was sent to school at Winchester College, where he was elected as a scholar in 1581. He won a scholarship to New College, Oxford, going up in August 1586. There he quickly won a reputation as a classicist and a poet.

His aristocratic new friends at Oxford were accepting of him as one of their own. In the romantic outpouring of grief and loss after the death of Philip Sidney in 1586, Bastard was one of those invited to write a memorial poem. He also contributed to the volume of elegies written on the death of Anne Cecil de Vere in 1588. His memorial book poems were written in Latin, elevating their subjects, giving them the aura of myth – and elevating their writers too, showing off their classical skills, neo-Martials and Juvenals and Horaces of the Elizabethan age. These poems were a mark of favour, an announcement of Bastard's arrival, in society, into the world of wits. To be a wit, to possess wit, was the highest distinction an Englishman of these times could have, vaulting its possessor over class and cultural boundaries.

Bastard was admitted as a Perpetual Fellow of New College in 1589. He received his BA in 1590, and everything seemed assured. Fortune was smiling upon him. His future was glorious.

But, as Wittgenstein said, the events of the future cannot be inferred from those of the present. Time does not move us in straight lines.

Abruptly, in 1591, Bastard's life changed, cataclysmically. An anonymous pamphlet lampooning the vices and liaisons of some Oxford clergy and academics and worthies (*An Admonition to the City of Oxford: Or his Libel entit. Marprelates Basterdine... exposing the amours of the University and town of Oxford*) was being distributed in the city. The pamphlet named names, locations, described adulteries, out-of-wedlock births, in brisk facetious rhymes,

> But see what tis to be un-Rulye
> He stroake his oare so deepe in water,
> That with the stroake, I tell you trewly,
> Her Belly swelld a Twelvemonth after

He would always deny his authorship, and maybe he took the fall for a higher-born friend (although that 'Basterdine' in the title does seem to be a clue), but Bastard was declared to be responsible. His fellowship was withdrawn and Bastard was expelled.

His friends didn't quite abandon him. Thomas Howard, the Earl of Suffolk, made him his chaplain, and then in 1592 presented him with a living as the vicar of Bere Regis back in Dorset. The living wasn't sufficient. After being cast out of Oxford, nothing would ever be sufficient; he chafed at his fall and exile and his fortunes would only get worse.

The churchwardens of Bere Regis sent the Dean a letter of complaint in 1597,

> Our Vicar doth not perambulate yearly... The bounds of the churchyard are in decay, in default of the Vicar... the pavement of the church is in decay to the sum of five shillings in default of the whole parish... our Vicar doth not read the Injunctions quarterly... there is a schoolmaster who teaches here, whether he be licensed or not we know not... There is a custom of cheese and beer to be paid of our Vicar on Christmas Day at night which he doth withhold from us.

And again, in 1600,

> We present that our children are not baptised, the fault is in our parson... the church is in decay... the glass in the church windows is still in decay...

What had Bastard been doing while his church decayed, while he neglected his duties and withheld Christmas beer and cheese from his parishioners and failed to baptise their children? He married three times, he accumulated his own children and a lot of debts; he wrote poetry, and complained about his grievances. His single uncomplicated pleasure seemed to have been for fishing.

It's all there in the book of epigrams, *Chrestoleros*, which Bastard published in 1598. Bastard suffered much. And, unstoically, uncynically, wrote about how much, and how wrongfully, he suffered.

He writes about his 'fatall fall', he complains about his wretchedness in the second person ('Thou seest thou art

forlorne') as well as the first ('smaller than I am I cannot be').
He repeatedly complains about his poverty.

> But nowe left naked of prosperitie,
> And subject unto bitter injurie,
> So poore of sense, so bare of wit I am,
> Not neede herself can drive an epigram.

But it's not just poverty and wretchedness that his epigrams
show, and the pains of trying to make something shine
beyond the efforts of its own making. They build into a sort
of cosmic resentment, a totalising, universalising bitterness.
A lament against the despoiling of nature by mankind ('We
have defaced the lasting monuments / And caused all honour
to have end with us') finishes with,

> What can our heirs inherit but our curse?
> The world must end, for men are so accurst;
> Unless God end it sooner, they will first.

Even his poem extolling fishing ('the sweet'st of sports')
concludes with complaint,

> But now the sport is marde, and wott ye why!
> Fishes decrease and fishers multiply!

Crucially, in book 2 of *Chrestoleros* is epigram 35, addressed
to Fortune, 'In Fortunam',

> I pray thee fortune, (fortune if thou be.)
> Come heere aside, for I must braule with thee.
> I'st you that sitt as Queene in throne so hye,
> In spite of vertue, witt and honesty?

41

Have you a Scepter onely to this ende,
To make him rue which never did offend?
I'st your fayre face whose favour fooles doe finde,
And whose vaine smile makes wise men change their
    mind?
Th'arte full of mosse, and yet a rolling stone.
Thou fancyest none: yet put'st the worste in trust,
Thou ta'kt no bribes, and yet dost judge unjust.
Thou makest Lordes, and yet dost cast them downe,
Thou hatest kings, and yet dost keepe their crowne,
Thou never stand'st: and yet dost never fall;
And car'st for none, and yet hast rule of all.
    But fortune, though in princely throne thou sit,
    I envie not, it is not for thy witt.

While continuing to harp on about his unjust expulsion ('To make him rue which never did offend'), in line 4 Bastard presents oppositions to Fortune. These are the weapons that he might use in his 'braule': virtue, wit and honesty. In the final line Bastard returns to wit, praising it by implication as the highest of all distinctions. As Philip Sidney wrote in his *Defence of Poesy* (written around 1580, but published in 1595), the poet has 'no law but wit'. Shakespeare gives us the same credo in *As You Like It*, written the year after *Chrestoleros* was published, 'Nature hath given us wit to flout at Fortune'.

What about wit? What about virtue? What about honesty? What about fortune? What, as Wittgenstein might have said, did Bastard *mean* by these words?

*

42

Bastard's thought and language were vehicles for images and meanings from the classical and Christian worlds.

In the Greek beginning, the principal gods cast lots to divide the universe. Zeus won the heavens, Poseidon the sea and Hades the underworld. A male fertility spirit, the Agathos Daemon, who was honoured in the form of a snake, was sometimes identified with Zeus. He acquired a consort, Agathe Tyche, Good Fortune. Their progeny was Plutus, Wealth.

Agathos diminished into a good-luck spirit that could be invoked to protect vineyards and wheat fields from pests and adverse weather. Tyche's power grew to command a realm where purpose, divine or human, does not operate. She became a city goddess, jealous of goodness, mightier than hope, an incalculable divinity, committing every crime, utterly unjust, bitter; it is impossible to escape from her and her 'gifts'.

Temples persisted to Tyche and to her Roman near-equivalent Fortuna into the Christian age. The bronze statue of Tyche of Antioch by Eutychides, which became the prototype for subsequent images, survives in replicas in the Vatican. The goddess is seated on a rock; in her right hand she bears a sheaf of wheat, symbolising prosperity and fecundity; on her head she wears the turreted crown whose battlements stand for the protection of the city; beneath her right foot is a young man.

Similarly, the Roman Fortuna, Jupiter's first-born daughter, was originally a goddess of fertility, whose name probably derives from *fors*, 'that which is brought'. She was an oracle-goddess, frequently portrayed with a cornucopia as the bestower of prosperity, and a rudder as the controller of destinies and fates, and with a wheel, or standing on a sphere, to indicate the uncertainty and mutability of life. As Machiavelli pointed out, the Romans built more temples to Fortuna than to any other divinity.

In the images of the rudder and the wheel, there is the opposition between Fate, which is fixed, and Fortune, which turns. No one ever prays to fate; there would be no point. But Fortune's decisions might be interpretable by oracles and responsive to prayers to intercede on our behalf.

Fortune was worshipped wherever life was most uncertain – the sea, the harvest, the sporting arena – in temples and shrines that persisted long after Christianity had taken root. Even when the image of the goddess had receded, the idea of fortune would remain gendered, Dame Fortune, personified always as a woman.

The concept, and images, of Fortune directly inherited by Bastard were provided by Boethius, the sixth-century Roman imperial administrator and philosopher who wrote his last work in prison while awaiting execution. In Boethius's *The Consolation of Philosophy*, Lady Philosophy appears in the condemned man's cell to instruct him in the correct philosophical attitude,

> Refusal to bear with your lot would make it more bitter since you cannot change it. Suppose you spread your sails before the winds; your course would then be dictated not by your own inclination, but by the direction from which they blew... Having entrusted yourself to Fortune's dominion, you must conform to your mistress's ways. What, are you trying to halt the motion of her whirling wheel? Dimmest of fools that you are, you must realise that if the wheel stops turning, it ceases to be the course of chance.

There is still something pagan here. The orthodox monotheist view is that God knows everything, it is all part of his plan, but here, we are all subject to the puff of fortune's wind, the turn of her wheel. Things happen. The word 'chance' (like the

Spanish *casualidad* and Italian *caso*) has its root in the Latin *cadere*, what falls. This is opposed to Fate, which comes from *fatum*, the gods' sentence.

Guides, from classical times on, instruct us in how to meet disappointments and disasters. The Duke Guidobaldi, in Castiglione's *Book of the Courtier*, suffers disastrous health and his enterprises fail, yet 'he always bore out with such stoutness of courage that Virtue never yielded to Fortune, but with a bold stomach, despising her storms, lived with great dignity and estimation among all men in sickness as one that was sound, and in adversity as one that was most fortunate'.

This attitude, the 'bold stomach' of virtue that Bastard never found, is derived from Seneca and the Stoics, and stretches back to Aristotle,

> for he is happy and blessed, not by reason of external goods, but in himself by reason of his own nature. And herein of necessity lies the difference between good fortune and happiness: for external goods come of themselves, and chance is the author of them, but no one is just or temperate through chance.

Boethius wrote about this too,

> virtue [*virtus*] is so-called because it relies on its strength [*vires*] not to be overcome by adversity. Those of you who are in the course of attaining virtue have not travelled this road merely to wallow in luxury or to languish in pleasure. You join battle keenly in mind with every kind of fortune, to ensure that when it is harsh it does not overthrow you, or when it is pleasant it does not corrupt you.

Virtue is an active stuff like the New Zealand *mana* or the Spanish *duende*. It has its root in the Latin word for man (*vir*): how a person should be, how the human essence is revealed, tested while suffering adversity or enjoying good fortune. This is how we find out who we are, and in the process escape from external determination. Petrarch had written about it, 'Many times whom fortune has made bond, virtue has made free'. The sixteenth-century notion of virtue contained elements of energy, boldness, generosity, will, courage, intelligence and resourcefulness.

✳

Bastard was a cleric, an Anglican vicar, therefore all things were God's to dispose, everything was part of the divine pattern that we are too limited, in our vision, in our understanding, from the low place where we look at the world, to apprehend and comprehend. As Alexander Pope wrote a century later,

All Nature is but Art, unknown to thee;
All Chance, Direction, which thou canst not see;
All Discord, Harmony not understood

Bastard was also a classicist, trained in the Latin models, schooled in Roman thought. The epigram form Bastard was using derives from the Roman poet Martial – Bastard's supposed friend John Harington wrote, 'It is certain that of all poems the epigram is the wittiest; and of all that write epigrams, Martial is counted the pleasantest.' The rules of rhetoric were from Cicero. Wit in this context was the display of *ingenium*, which meant ornamentation and comparison and metaphor. So that, for epigrammatists and playwrights, 'wit' and 'ingenuity' were almost synonymous. Bastard's entry to the world of well-born wits at Oxford had been secured by poetry. (And wit, whose root is Old English, meaning 'to know', also had connotations of male sexuality – as the very expression of what a man was, 'wit' might sometimes be used to signify the penis.)

And here I sit now in the British Library, reading the lives, and unfortunate deaths, of Elizabethan wits, not quite remembering how I ended up keeping such close company with Thomas Bastard. Probably, I happened upon his ode to Fortune, and the poet's name had attracted me in a slightly schoolboyish way to investigate his life story. He has become an emblematic figure for me – *poor Bastard*, the embodiment of the victim of luck whose shoulder Tyche's foot presses down upon.

I'm going through Bastard's volume of epigrams, the publication of which he maybe anticipated with hope and expectation, allowing himself to believe that he might be lifted again by the wheel, and his casual challenge to fortune is just a light-hearted self-mockery, not really a challenge at all, because everything once had been going so well and maybe it could again...?

Chrestoleros (Bastard's own Greek neologism, suggesting both useful and trifling) was not greeted with favour. In it Bastard claims to have 'taught epigrams to speak chastely, barring them of their old liberty, not only forbidding them to be personal, but turning all their bitterness rather into sharpness'.

Many disagreed. As John Harington reported, 'the dusty wits of this ungratefull time / Carpe at thy booke of Epigrams, and scoffe it.' This sounds like the supposedly well-intentioned thing a malicious friend offers in the aftermath of a failure: *I liked it well enough, but you know how people are, they don't appreciate truly good things...*

In an epigram of his own, Harington accused Bastard of flattery and cowardice,

Then what's the reason, Bastard, why thy rhymes
Magnify magistrates, yet taunt the times?
I think that he to taunt the times that spares not
Would touch the magistrates, save that he dares not.

*

Presumably if Bastard was, as he always claimed to be, innocent of the Oxford libel, then he would have known who its real author was. In an unvirtuous transaction, Bastard might have traded his silence for an insufficient clerical living. *Chrestoleros* includes craven flatteries of eminent men – the Archbishop of Canterbury, local landowners – in between plaints about his fallen state. At no point does he seem to indicate a guilty man, only persists in declaring his own innocence and the injustice done to him.

Harington wasn't beyond being a flatterer himself, constantly seeking patronage and place from his godmother, Elizabeth, and then from her successor, James I. Banished by the queen for his racy translation of canto 28 of Orlando's *Ariosto Furioso*, in his exile he translated the entire poem and invented the flush toilet, which he called the Ajax. Forgiven his earlier scurrility, he was invited back into favour, but his *A New Discourse of a Stale Subject, called The Metamorphosis of Ajax* had him banished again. Harington would experience the turns of fortune's wheel repeatedly. It helped that he had a family estate to fall to.

Harington aimed higher than Bastard, confident that his wit could lift him from his falls. He installed a flush toilet at Elizabeth's palace at Richmond. The queen summoned him to entertain her on her deathbed.

As one 'dusty wit' wrote to another, 'Bastard has the name of a very good lively wit but it does not lie this way in his epigrams... [he] botches up his verse with variations, and his conceits so run upon his poverty that his wit is rather to be pitied than commended.'

Botches up his verse with variations... His attempts at wit, at ornamentation, metaphor, *ingenium*, were failures. Those lines in his Fortune poem about rolling stones and moss would have been tired and clichéd even in his day.

Bastard himself said that his wit had become 'key-cold'. It takes confidence to produce epigrams. The writer stands apart from, and at least slightly above, the follies he describes. When he cowers below his subject matter, the effect becomes that of resentment forcing itself to be jolly, which is not a comfortable position or tone.

John Hoskins (sometimes Hoskyns) is a useful comparison here. Hoskins was a contemporary of Bastard's who was also from the yeomanry; a fellow scholar at Winchester and New College, he too had his fellowship withdrawn, being expelled shortly after Bastard, for 'insolence'. Hoskins had also contributed to the 1587 Sidney memorial volume – in the contemporary affectation of adopting Latinised versions of names they are listed there as Joannes Hoschines and Thomas Bastardvs. And they were convicted of the same crime, being rude about people with power and authority.

Hoskins was 'the flower of his time', according to the diarist John Aubrey, 'but he was so bitterly satyrical that he was expelled and putt to his shifts'. After his expulsion Hoskins taught for a year in Somerset and then married a rich widow, after which he became a lawyer and was admitted to the Middle Temple in London. 'He wore good cloathes, and kept good company. His excellent witt gave him letters of commendacion to all ingeniose persons… His great witt quickly made him be taken notice of.'

While Hoskins thrived, Bastard longed for the metropolis but remained in Dorset, where he wrote effortful epigrams and dull sermons and shirked his job, idling away his time in bitterness and resentment and fishing.

Bastard had been delighted to hear that Hoskins had followed him in being cast out of Oxford,

When my friend Candidus was in distresse,
Methought I joyed true felicitie.
To love his woe it was my happinesse,
And to feel halfe of my friend's misery.
But when his fortune turned about her wheele,
And melancholy good did overtake him,
I was no fitt companion for his weale.
From thence began my woe and my forsaking.
   For now he keeps the good as cruellie
   As franke of late he spent the evill on me.

If Bastard failed at wit and virtue, this does at least sound like honesty, a candour about his own resentments, his pettiness of feeling. But honesty, in Bastard's contemporary context, doesn't equate to frankness or self-exposure. Honesty was the antonym to 'knavery' and 'flattery', meaning not so much emotional candour, but truth-telling and plain-speaking to power, which Bastard, after the Oxford libel, was seldom capable of.

Hoskins, like Bastard, had no estate or family money. He used his advantageous marriage well. Bastard married three times, 'the first was joined to me at a tender age for love, / the second for riches, the third for comfort'. He wasn't able to hold on to any of those states, or use them, or his wit, to advance himself in the world.

There are alleviated moments, and not just to do with fishing. Bastard writes an unforcedly charming ode to a child learning to speak,

> Methinks 'tis pretty sport to hear a child,
> Rocking a word in mouth yet undefiled
> ...
> And the soft air the softer roof doth kiss
> With a sweet dying and a pretty miss

Hoskins also had a poem on infant speech, but works it as a caution against following his own example,

> My little Ben, whil'st thou art young,
> And know'st not how to rule thy tongue,
> Make it thy slave whil'st thou art free,
> Lest it, as mine, imprison thee.

Sent to the Tower of London in 1614, for being again too outspoken, this time in the House of Commons, Hoskins was freed after a year, and flourished. He lived well into his seventies, until, as Aubrey reports, while at assizes at Hereford, 'a massive countrey fellowe trod on his toe, which caused a gangrene which was the cause of his death'.

Bastard wrote,

> Had I my wish, contented I should be,
> Though neither rich nor better than you see,
> For 'tis not wealth nor honour that I crave,
> But a short life, reader, and a long grave.

This much he did achieve.

He took on a second parish, in nearby Almer, in 1606, but his fortune still didn't improve. Nor did he exercise his virtue or prudence to improve it. His predecessor at Bere Regis had rented out the church lands to add to his income. There is no record of Bastard having done so.

Why was the scandal of 1591 such an irrevocable moment for Thomas Bastard? Why did he decide that being expelled from Oxford was too big to fight against? The confidence needed to write epigrams was lost to him, so the exercise became a forlorn attempt to prove his wit and skill.

For a time Bastard had some influential friends – and *confidence*, the moment flowed through him… But maybe he was a victim of good fortune rather than bad.

Boethius wrote, 'My opinion in fact is that adverse Fortune benefits people more than good, for whereas when good Fortune seems to fawn on us, she invariably deceives us with the appearance of happiness, adverse Fortune is always truthful…'

In 1611, Bastard contributed to a volume of burlesque poetical tributes to the traveller Thomas Coryat called *Coryat's Crudities*, among whose other contributors were, inevitably, John Harington and John Hoskins. It must have been a reminder of his younger, more glorious days, when Bastard was in the company of other poets and wits, when his star was in the perpetual ascendant, before he challenged fortune, before fortune turned her back. His only other publications were two collections of sermons and a panegyric in three volumes to the new king, James I (*Serenissimo potentissimoque monarchae Iacobo Magnae Britanniae, Franciae, & Hiberniae, regi Magnam Britanniam*), which the new king ignored.

Despite his resentments and his poverty, Bastard continued, despite himself, to hope. The flatteries that Harington mocked him for (and which Harington himself was even guiltier of, albeit on a grander scale – always looking for a patron, a place, a preferment), they were aspects of hope: the unfortunate who had felt fortune's wheel grind him down was still longing for the intercession of something stronger, that he didn't know what to call, to lift him up again.

Bastard would not have achieved his scholarships without a facility in the classics; he demonstrably had an equivalent one in versifying, but maybe, in fact, fortune had favoured him too much at the beginning, and that was what did for him, lifting him in ambition and society to a place that his talent – his wit, his virtue, his honesty – wasn't large enough to hold.

But there is still something missing, exemplified by the manner of Hoskins's death.

How can Fortune contain both necessity and chance? It's the implacable wheel that turns – and all those other mechanistic metaphors going back to Roman times, the ship's rudder and sail – and it's the unpredictable event: the country fellow stepping back to trample on Hoskins's toe; Aristotle's gardener finding buried treasure; or Saul in the Old Testament, the shepherd boy who goes looking for his father's lost asses, and wanders into becoming a king.

The unpredictable chanceful event that falls where it might, that can be withstood, maybe even overcome by human attributes of virtue and wit and honesty, needed another word to define and describe it. The ancient divinities were almost gone; the Christian god was receding from the universe, and the gap he left was getting bigger.

That word already existed but it appears nowhere in Bastard's poetry. Harington wrote about it in his 'Treatise on Playe', an unpublished account of what we would now term 'problem' or 'addicted' gambling, and which he called his 'infection' to 'deep play'. Shakespeare used it often.

It was a gambler's word. Fifteenth-century English gamblers had invented it, adapting the Middle High German *gelücke* (in modern German, *Glück*), which means both happiness and good fortune, to designate that capricious force that intervenes between desire and its consummation, to account for the disparity between merit and consequence, taking God and fate and destiny out of the process, desacralising it all, and they called it Luck.

The idea of luck contained then, and still does, many of the attributes formerly ascribed to Fortune. 'Lady Luck' and 'Dame Fortune' share a gender. They're both capricious, heartless, cruel. The benefits they give or penalties they exact are irrespective of merit ('unearned luck', as Puck says in *A Midsummer Night's Dream*). The language of advice given by experts and philosophers of how to maximise one's luck is often violent – how to seize 'her' – with connotations, or in Machiavelli's case, overt instructions, of sexual violence ('fortune is a woman, and if you wish to keep her under, it is necessary to beat and ill-use her').

Like justice, or its mother the Roman goddess Fortuna, luck is often conventionally depicted as blind. My father, when he could be persuaded to talk about his early years, would refer to luck, at least his, as 'dumb'. It's an American usage, 'dumb luck', which of course denotes stupidity, but it also adds to the sensory deprivation: luck cannot see, nor may it speak. Neither does it have a memory: toss a coin ninety-nine times and each time it lands heads; on the hundredth toss, the odds are still 50:50 that it lands heads or tails.

Luck can, like a horse, or a bicycle, be ridden, with the implication being that you're likely, as Boethius did, and Harington, and Hoskins, and Bastard, to fall.

*

Anthony à Wood, the seventeenth-century antiquary who devoted himself to writing the biographies of Oxford-educated writers, closed Thomas Bastard's with this summary, 'This poet and preacher being towards his latter end crazed, and thereupon brought into debt, was at length committed to the prison in Allhallows parish, in Dorchester, where dying very obscurely and in a mean condition, was buried in the churchyard belonging to the parish on 19th April 1618, leaving behind him many memorials of his wit and drollerie.'

The physical estate Bastard left consisted of various household goods: five beds, four chairs, two blankets, some kitchen dishes and utensils. The only clothes were 'Item one gowne a paier of breeches and a dooblett', which would have comprised his prison wardrobe. Presumably his shoes went to his jailors. Making up nearly half the value of the estate were 133 books in a chest.

This book is in part a gallery of portraits of luck's heroes to emulate and luck's martyrs whose examples it would be imprudent to follow. But some things are not known. I could tell you that his brother John administered his estate, that his scholarship to New College began on 27 August 1586. What I can't tell you is what Thomas Bastard looked like, the effect of his physical presence in a room. His appearance is never mentioned, whether he was well- or ill-favoured, because 'wit' was the thing, but I imagine he had a personal beauty, the sort that attracts others, that makes his end of the table the place to be.

There are paintings extant of John Harington. I've tracked down a couple of portraits of John Hoskins, but I haven't been able to find any representation of Thomas Bastard. I imagine that when he was young, when fortune, and confidence, and luck, were still with him, there must have been an excitement surrounding him, a physical ease and charm to go with his mental dexterity. 'He was a person endowed with many rare gifts,' wrote Wood.

If you search for Thomas Bastard online, there's his wiki page, and an entry by the Poetry Foundation, which gives a summary of the life (Dorset, Winchester, New College, the Oxford libel, his clerical career). It tells us, maybe a little generously, about his *Chrestoleros: Seven Books of Epigrames* (1598) – 'These brief poems, ranging in length from two to 16 lines, are primarily concerned with the events and people of his time and balance lively satire against bitter reflections of poverty.' It concludes, 'After a mental breakdown, he died at the age of 52 in a debtor's prison in Dorchester and was buried in a churchyard there.' It offers links to nineteen of his epigrams, although not the one on Fortune.

Further down, there's a print-on-demand Amazon link, that charming epigram on infant's speech, clearly observed from life (all those beds left behind) and chosen as a featured poem by a London community worker; you go down that page, click on to others, and there are gliding mentions of Bastard in books digitised by Google (*The English Poetic Epitaph*; *Verse Libel in Renaissance England and Scotland*; *The History of English Poetry*).

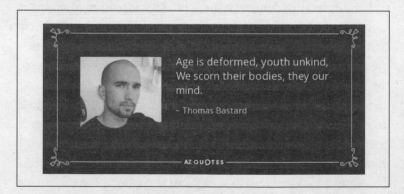

Age is deformed, youth unkind,
We scorn their bodies, they our mind.

~ Thomas Bastard

AZ QUOTES

And then you get the mistaken identities. The site Goodreads adds to his bibliography *The Autobiography of Cockney Tom* ('Showing his Struggles through Life, and proving this Truth of the Old Saying "that Honesty is the best Policy"'). It's set in England, South Australia and the goldfields of Victoria, Australia, in the 1800s and was written by a much later Thomas Bastard.

There is one picture that persistently pops up, in different contexts, claiming to be of him.

This is not our Bastard, against whom fortune's spite continues to exert itself. The photograph is of a French boxer of the same name, who fought intermittently between 2003 and 2017, winning thirteen fights and losing three.

*horse racing –£10**

---

* Tyche finished fourth. According to the *Sporting Life*, she 'ran on strongly final furlong but never reached challenging position'.

CHAPTER 4

Roulettenburg

Some people have luck, and everything comes out right
with them; others have none, and never a thing turns
out fortunately.

Fyodor Dostoevsky, *The Idiot*

It is a strange thing, I have not won yet, but I am
acting, feeling, and thinking like a rich man, and I
cannot envisage anything else.

Fyodor Dostoevsky, *The Gambler*

net total −£10

I had not felt the need to visit Blandford Forum where
Thomas Bastard was born, or Dorchester, where he died. I
had been satisfied that I'd 'got' what Wittgenstein meant by
luck without being at the summer fair, or Thomas Bastard
by wit and Fortune, by reading their words and others'. But
there are times when the library won't suffice. The more I
read about, and by, Dostoevsky, and his relationship to luck,
the further he receded from my understanding.

There was a sickly feeling in all this that I didn't quite
recognise. It might have been close to what Sigmund Freud
was writing about to his student Theodor Reik, 'You are

right, too, in suspecting that, in spite of all my admiration for Dostoevsky's intensity and pre-eminence, I do not really like him.' Dostoevsky gambled intermittently and frenziedly throughout much of his adult life: as a military cadet; through his first marriage and its aftermath, in the company of 'the electric woman' Polina Suslova. Even in the early years of his redemptive second marriage, he would leave his much younger wife behind while he gambled away money he couldn't afford to lose in the spas of Germany. The morning after another mad night of losses, when he had failed, again, to maintain his emotional composure, he'd abase himself, promise to give it up. Then, somehow briefly purged, he'd write the next section of *Crime and Punishment*, borrow some money from a well-wisher whom he'd now detest for having helped him, and go back to the casino, convinced he'd finally set right his own fortunes, and his wife's, and the fortunes of all those who depended on him, such as his stepson, and the children and wife of his dead brother Mikhail, and do it all over again.

Dostoevsky's gambling had been interrupted by his 1849 arrest for his involvement in what was essentially a progressive discussion group, and his subsequent exile, prison sentence in Siberia, and army service. He returned to St Petersburg in 1859, his politics moving towards a reactionary Christian nativism, whereupon he resumed his masochistic relationships with hope and loss.

I've seen compulsive gamblers. I recognise their submission to luck, the self-defeating way they spend money as if it is already theirs, when everything that has gone before shows that it is money they will never possess. I understand that, despite their grandiosity, their airy scheme-making, their violently anti-inductive position, they expect to lose, that, probably, they want to.

Freud diagnosed Dostoevsky as an emotional masochist, 'For him gambling was a method of self-punishment... the one thing which offered any real hope of salvation – his literary production – never went better than when they had lost everything and pawned their last possessions.' He also suggests that Dostoevsky's epilepsy was a psychic symptom rather than neurological, and that his addiction to gambling was a replacement for 'the vice of masturbation', Oedipally determined by a fearsome father and a cosseting mother. Whether this convinces or not, it doesn't explain why Dostoevsky finally succeeded in giving up gambling in Wiesbaden in 1871.

Casinos were illegal in Russia, so Russian gamblers would go to the spa towns of Germany to play. In a letter from Paris to his first wife's sister in September 1863, Dostoevsky, typically, wrote of his recent experiences in Wiesbaden,

Varvara Dmitrievna, I watched those players attentively for four days. There are several hundred gambling here and, to be honest with you, only two of them really know what they're doing. They all lose heavily because they don't know how to gamble. There was a Frenchwoman and an English lord. They knew how to play. Not only didn't they lose but they almost broke the bank. Please don't think that I'm bragging about the fact that I didn't lose in saying that I know the secret of not losing, but winning. I really do know the secret. It's very stupid and simple and amounts to ceaseless self-control at all stages of the game and not getting excited. That's all there is to it. That way you can't lose and are bound to win. But that's not the point. The point is whether, once you know the secret, you are capable of exploiting it... But on to business.

He then details his arrangements for the disbursal of his winnings: some to his brother, Mikhail, some to his wife, Maria. A week later, from Baden-Baden, which he would rename 'Roulettenburg' for his novel, *The Gambler*, he wrote again, asking for 'a favour'. He'd lost almost everything he had, so could she arrange for his wife to send him back some of the money he'd sent home?

He had intended to travel on to Rome, instead he became stuck in Turin, unable to pay his bills. In a letter to his brother, he asserts that his pains at having to admit to his losses are at least as great as Mikhail's at having to subsidise them, again, even if Mikhail is ill and short of funds and trying to keep their co-edited literary magazine running. Fyodor then asks for more money and relates what had happened in Germany,

I invented a system of playing, I put it into practice and won at once 10,000 francs. Next morning, in excitement, I was unfaithful to my system, and lost at once. In the evening I returned to my system again, in full rigour, and immediately and without any effort I won 3,000 francs. Tell me, after that, how could I help being carried away, how could I help believing that if I followed that system, luck was in my hands? And I need money, for myself, for you, for my wife, for the writing of my novel. Here scores of thousands are won as though it were a mere joke... Then I suddenly began losing, could no longer restrain myself, and lost everything to the very last. After I had posted my letter to you from Baden, I took *all the money* I had left and went to the tables: with four napoleons I won 35 in half an hour. The extraordinary luck bewitched me; I risked the thirty-five and lost them all. After paying the landlady we were left with six napoleons for our journey. At Geneva I pawned my watch.

A rush of words follows, self-exculpating, self-incriminating, bringing the correspondent close to the writer, pushing him further away, asserting the lowliness of Dostoevsky's position, the misery he was enduring, his desperateness. In Baden-Baden he'd gone to see Turgenev. He'd borrowed money from the richer writer, although he does not tell his brother that. Instead he complains of Turgenev's sulking and being made to listen to his 'moral sufferings and doubts', which, presumably, was the price on the loan.

He gave me his 'Ghosts' to read, and owing to my playing I could not manage to read it, so I returned it to him without reading it. He says he wrote it for our review, and that, if I write him from Rome, he will send it to me there. But what do I know about the review?

Do not say anything about my position to any one. It is a secret, I mean, my losses.

I am in a great hurry to get out of beastly Turin. And I have still a good many letters to write; to Maria Dmitrievna and to Varvara Dmitrievna.

Give my thanks to Varvara Dmitrievna. What a glorious soul! This is what I am afraid of. I am afraid that Maria Dmitrievna may write you something unpleasant. Yet I do not suppose she will. She may, of course, need no money till the middle of October. But how can I tell? Perhaps I have placed her in a false position. She needed a hundred roubles for something, but had not decided to make that expense. But after getting my letter, in which I said that I was sending her money, she may have incurred that expense. And now she is perhaps without money. I tremble at the thought of it. If only someone had sent me an account of her health!

Maria died the following year. So did Mikhail. Their review, the *Vremva*, had already been closed by the censor. Dostoevsky then assumed responsibility for his brother's debts and dependants as well as his own.

He reached out for loans wherever he could: to Turgenev, to Wrangel ('my excellent and old friend'), to Katkov, to Herzen, to Lubyimov ('with sincere respect and devotion'). His letters were like those of a particular type of his characters: ardent, self-obsessed, driven by circumstance and weakness to flatter lesser men. He was still in love with 'the electric woman' Polina, who pawned her jewellery for him and permitted him her company, but no longer anything beyond that. (She had recently 'surrendered her heart' to a Spanish medical student named Salvador, who was much closer to her age but had not offered his own heart in return.)

In June 1865 he made a successful application to the Society for Assistance to Needy Writers and Scholars, but money kept pouring away from him.

Dostoevsky had been thinking of writing a gambling novel since at least 1859, when he read the account in *Russkoye Slovoe* of a young gambler in Wiesbaden who had won a fortune at roulette only to lose it all soon afterwards. He 'shot himself there in the casino before the eyes of numerous spectators, crowding around the roulette table. It is noteworthy that this sad event did not even interrupt the gambling. The man calling out the numbers went on turning the wheel with the same cold-bloodedness with which he ordered the attendant to clean off the green baize of the treacherous table onto which the brains had spurted from the dead gambler's shattered head.'

He wrote about his as yet unwritten book in a letter to a friend in 1862,

I have in mind a man who is straightforward, highly cultivated, and yet in every respect unfinished, a man who has lost his faith but who does not dare not to believe, and who rebels against the established order and yet fears it... The main thing, though, is that all his vital sap, his energies, rebellion, daring, have been channelled into roulette.

By the summer of 1865, the book remained unwritten, but Dostoevsky signed a contract with the most notoriously unscrupulous publisher in Russia, Fyodor Timofeyevich Stellovsky. Stellovsky paid Dostoevsky 3,000 roubles for the right to print a complete edition of Dostoevsky's works. In addition, Dostoevsky would deliver his new novel by 1 November 1866. The novel should be in seven large-size folios, on two-columned pages; if he failed to deliver, Stellovsky would have the right to publish anything Dostoevsky wrote for the next nine years, without payment.

His gambling novel was a book entire, ready-made. It had been in his thoughts for years. Once he had actually turned to write it, he was sure that it would, in that feverish way that writers often rise into when considering a new project, write itself. Dostoevsky would barely even need to be in the same room. And his money worries would be on their way to being solved.

Dostoevsky travelled again to Germany, where he lost money on roulette, borrowed more from men whose generosity made the writer despise them, returned to St Petersburg, and resumed work on *Crime and Punishment*, which was already contracted elsewhere. Every so often he would try to attend to the promised, unnamed, novel, but nothing was coming. He toyed with the notion of a group novel, enlisting the assistance of literary friends, each of whom could write a few chapters under his direction, but that too fizzled into nothing.

By October 1866, with less than a month to go before the ruinous deadline, he still hadn't written the novel for Stellovsky.

It was recommended to him that he should try dictating. Maybe that would free him up, and the material. So he hired a young stenographer, Anna Grigorevna Snitkina. She was twenty, he was forty-five, her father's favourite writer. In her *Reminiscences*, she writes of their first meeting, her shorthand teacher having sent his best pupil to Dostoevsky's apartment, with the warning, 'I am only afraid you won't make friends with him; he is such a gloomy, stern man.'

She describes the study they were to work in, the brown and red fabrics covering the armchairs and table. She remarks on the writer's appearance, his chestnut-coloured hair, heavily pomaded and tinged faintly with red. He seemed very old to her, with a very erect posture. She was particularly struck by his eyes – one was dark brown, the other had no visible iris at all, its pupil permanently dilated as a result of atropine treatment for an injury suffered in an epileptic fit.

Her position was at the writing desk at one end of the room. On the opposite wall, in a walnut frame, hung a portrait of an emaciated woman wearing a black dress and cap.

In five minutes the maid came in carrying two glasses of very strong, almost black tea. Two rolls lay on the tray. I took a glass. I didn't really want it and the room was already overheated, but I drank it anyway, in order not to appear fussy...

At about four o'clock I got ready to leave, promising to return with the finished work the next day at noon. At parting he handed me a packet of heavy writing paper ruled almost invisibly, the kind he habitually used, and showed me exactly the margins I was to leave on it.

So our work began and went on. I would come to his house at twelve and stay until four. During that time we would have three dictating sessions of a half-hour or more, and between dictations we would drink tea and talk.

In one of those conversations between dictations, he told her he was on the point of choosing between three different paths: to travel to the East, to Constantinople and Jerusalem; to get married; or to go abroad to play roulette and be a gambler.

From 4–29 October, in twenty-six days, he dictated his new novel to her. Its story of Aleksy Ivanovich, a superior but impoverished tutor to a Russian family holidaying at a German spa, contained many of the elements of the recent episodes of his life: the feverish, compulsive gambling; social resentment; the constant rebuffs to a prickly man's pride dealt him by unworthy foreigners and fellow Russians who happen to have more money than him; the hero's abasement before the unavailable woman he adores, who in the novel, as in life, is called Polina. And the novel allows itself, the writer awards himself, a gambling triumph, a climax that Freud might have seen as vindication of his masturbatory analysis. Polina has come to Aleksy's room finally to offer her love. Instead of consummating the moment with the woman he adores, Aleksy returns to the casino, where he follows his, and Dostoevsky's, customary procedure of staking everything on repeated 50:50 or 1-in-3 bets, and is rewarded with a huge fortune, which he will fritter away, just as he has frittered away Polina's love for him.

Anna presented the completed manuscript to Dostoevsky on 30 October, his birthday. He would make the final corrections and deliver the novel to Stellovsky's office on 1 November. The writer invited the stenographer out for a victory dinner, but Anna refused, from shyness, because she had never been to a restaurant before.

Dostoevsky was nervous that Stellovsky 'would contrive some kind of trick... would find a pretext for refusing to accept the manuscript'. The publisher was indeed out of his office when Dostoevsky came to deliver *Roulettenburg*, and the manager of the publishing firm would not accept it, claiming he had insufficient authority to do so. On Anna Grigorevna's advice, Dostoevsky took the package to the nearest police station where, after hours of waiting, he received a dated receipt at 10 p.m. The novel, satisfying the requirements of their wager, freeing Dostoevsky from the worst consequences of their contract, was renamed by Stellovsky and published as *The Gambler*.

*

The book is the best account of the gambling psychology I know. It is a first-person narrative, ruthless in its depiction of the lies that addicts know they're telling themselves,

> I watched and took notes; it seemed to me that calculation in itself means little enough and has none of the importance many gamblers give it. They sit there with bits of paper all ruled out, noting down the wins, counting up, working out the chances, making calculations, finally placing a bet, and – losing in exactly the same way as we simple mortals, who play without any calculations. On the other hand, I did reach one conclusion which, it seems is correct: in a random series of spins there really is, if not a system, then some sort of order – which is, of course, extremely odd.

You will see these roulette players in every casino. They collate each turn of the wheel and dutifully write down the numbers that land or else they strike them out on the pre-printed cards with the pens that the casinos so helpfully provide, encouraging the idea of science and research and precedence and above all the possibility of finding a pattern, which is impossible if the roulette table is true – and by 'true', we mean 'random' (except maybe they're right: no roulette table is ever entirely true…).

'Something,' Aleksy writes, 'would happen to me in Roulettenburg, that there would be something, quite without fail, that would affect my destiny radically and definitively.'

This, the action of luck, will lift him or crush him – as the matriarch of the Russian family is crushed by her own roulette mania. Aunty is so struck by the amount that might be won by betting on 0 that no one can persuade her to do otherwise; she is rewarded with an early win, and then loses almost everything by continuing to do so, leaving her family without the prospects it had expected, returning to Russia to try to expiate herself by spending what is left of her money on building a church.

In the climactic gambling scene Aleksy puts his entire roll, twenty gold friedrichs, on *passe* (the high numbers, 19–36). The ball lands on 22. He lets the money stand: 31. Now he has eighty gold friedrichs. After two 50:50 bets, comes a 1-in-3 chance. He bets the entirety on *moyen douzaine* (the middle numbers, 13–24): 24.

They laid out three rolls of fifty gold friedrichs for me, and ten gold coins; in all, together with what I had before, I now had two hundred gold friedrichs.

Feeling delirious I moved the whole pile of money onto the red – and then I suddenly came to my senses! And for the only time during the whole of that evening, during the entire game, an icy fear came over me, making my hands and legs tremble. With horror I felt, and momentarily realised, what it would be like to lose now! My whole life was at stake!

'Rouge!' cried the croupier, and I drew breath…

Now he has 4,000 florins and 80 gold friedrichs. He stakes 2,000 on *moyen douzaine*, and loses. ('I was overcome with rage…') He stakes the remaining 2,000 florins on *premiere douzaine*: 4.

He has 6,000 florins. Stakes 4,000 on black: black. Loses the next three bets. Stakes the remaining 4,000 on *passe*: wins; and wins the next four times.

This is the climactic scene that he had dictated to Anna Grigorevna in that gloomy St Petersburg apartment presided over by the portrait of his malnourished first wife. It is impossible to suppose that it could have been done calmly. The passionate gambler would have been walking up and down, voice rising in excitement as the action builds. We can imagine shy Anna nervously guessing at certain words, maybe the French terminology for gambling procedures that were unfamiliar to her, because she didn't dare interrupt her father's favourite writer in his flow.

In half an hour, Aleksy wins 30,000 florins, which is more than the bank will bet against; so the table closes and he moves to another. 'I set about staking once again, haphazardly and without counting. I do not understand what saved me!'

He breaks the bank on this table as well, and is urged, first by 'some Jew or other from Frankfurt' and then by a woman 'with a rather unhealthily pale, weary face' to leave the casino; but he just wordlessly gives the lady a roll of notes of 50 gold friedrichs and moves on to trente-et-quarante. It is a game he doesn't understand, beyond that it is possible to bet on red or black.

The experienced player knows what 'capricious luck' means. For instance one would think that after red had come up sixteen times it is bound to be black on the seventeenth. Newcomers to the game fall for this in crowds, doubling and trebling their stakes and losing dreadfully.

But owing to some strange sort of waywardness, after I had noticed that red had come up seven times in a row, I deliberately stuck to it. I am convinced that half of it was vanity; I wanted to astonish the spectators by taking crazy risks, and – oh, what an odd sensation! – I distinctly remember that without even the slightest prompting from vanity, a frightful craving for risk suddenly took hold of me. It may be that by going through so many sensations the soul does not feel satisfaction but is only exasperated by them, and demands yet more sensations, and stronger and stronger ones, until it is finally exhausted. And I promise I am not lying when I say that if the rules of the game had allowed me to stake fifty thousand florins at a time, I would certainly have done so. All around people were shouting that it was a folly, that red had already come up fourteen times.

'Monsieur a gagné déjà cent mille florins,' someone's voice beside me rang out.

He leaves the casino, banknotes and gold crammed into his pockets.

Several hands stretched out towards me; I gave it away in handfuls, as much as I could get in my fist. Two Jews stopped me by the door.

'You are brave! You are very brave!' they said to me, 'but you must leave tomorrow morning without fail, otherwise you'll lose absolutely all of it...'

One might expect Dostoevsky to have cured himself of his own mania by writing it. His deal with Stellovsky had served its purpose. He had bought time to complete *Crime and Punishment*. And he had a further reason for stability: four months after their first meeting, he and Anna were married.

In April 1867, to give them some time together, away from the persistent attentions of the writer's dependants and creditors, and because Dostoevsky was convinced that the symptoms of his epilepsy were mitigated when he was in Europe, the couple went on what was intended to be a vacation of three months, which stretched into four years.

They took the train from St Petersburg to Berlin, and then to Dresden, where they rented an apartment in a private house. He went off on his own to cafés to read newspapers – French, if there were no Russian ones available. Anxiety about their debts followed them to Germany. Court action had been taken against Dostoevsky by two of his creditors. When they were alone together, the newly-weds bickered with each other over anything from the size of the King of Saxony's private guard, to sunsets, to Dostoevsky's skill as a fairground marksman. 'A wife was the natural enemy of her husband,' Dostoevsky told her. And the two of them, particularly the husband, bickered with the people around them. 'He seems to take a perfect delight in saying uncivil things to the Germans,' Anna wrote in her diary.

After three weeks of this, Dostoevsky left his wife behind in Dresden and took the train to the spa town of Bad Hombourg. He had resolved to deal with their debts in his traditional way.

Bad Hombourg 22 May 1867, ten o'clock in the morning, to Anna Gregorevna Dostoevsky,

Forgive me, my angel, for going into some details concerning my enterprise, concerning this game, so it will be clear to you what this is all about. Already some twenty times or so I have gone to the gaming tables and had the experience that if I play with sangfroid, *calmly*, and calculatingly, there is *absolutely no chance of losing*! I swear it to you, there is no such chance! There's blind chance, while I have a calculation on my side, consequently, the odds favour me. but what has usually happened? I have usually started play with *forty gulden*, took them from my pocket, sat down, and placed them, one or two gulden at a time. In a quarter of an hour, I have usually (*always*) won double. That would be the time to stop and go away, at least until the evening, so as to give my excited nerves a chance to calm down (besides, I have made the observation – most certainly correct – that I can be calm and composed for *no longer than half an hour at a time*, when I am gambling). But I would walk away just to smoke a cigarette, and then return to the gaming table immediately. Why did I do that knowing almost for certain that I would not contain myself, i.e., go on to lose? That's because every day, upon getting up in the morning, I told myself that this would my last day in Hombourg, that I would be leaving the next day, so that, consequently, I could not play a waiting game at roulette. I tried hurriedly, with all my strength, to win as much as possible, right away, that very day (for I was to leave on the following day), lost my sangfroid, my nerves got excited, I started taking chances, I got angry, proceeded to place my bets without any calculation, having lost the thread of it, and eventually lost (because anybody who plays without calculating, helter-skelter, is a madman). My whole mistake was to have parted with you, rather than taking you with me. Yes, yes, this is so.

And so his letters go, accounts of composure and money lost, self-recriminations for leaving her behind, the sustaining hope that he will somehow regain an emotional continence and win back the money they need, expressions of great love for his precious angel, the pain he feels when she writes to him of her misery, her loneliness, her inability to sleep, her tears, and his repeated use of the words 'calculate' and 'calculation', along with his reiterated belief that he knows the secret of successful gaming.

He would tell of watching someone who could execute this, who would be un-Russianly and inhumanly (the two are synonyms for him) composed, in charge of his emotions, victorious; quite often it would be a Jew playing 'with horrible, *inhuman* composure', raking in the money. His self-diagnosis is contrary to the one Freud would later make.

Goodbye, Anya, goodbye, my precious angel, I am terribly worried about you, but you have absolutely no need to worry about me. My health is *excellent*. That nervous disorder you fear in me is only physical and mechanical! It is not a mental perturbation, that is for sure. It is something that my nature demands, it is how I am made. I am nervous and I could never be calm even without all this! Besides, the air is wonderful here. I *couldn't* be healthier, but am really suffering for you. I love you and that is why I am suffering.

I hug you hard and kiss you countless times,
Yours,
F.D.

This is the sixth daily letter he'd sent since leaving her behind in Dresden. She wrote in her diary,

> What can I do? Apparently he just has to. The best thing will be if that unhappy notion about winning gets knocked out of his head.

Later she would write,

> I loved Fyodor Mikhailovich without limit, but this was not a physical love, not a passion which might have existed between persons of equal age. My love was entirely cerebral, it was an idea existing in my head. It was more like adoration and reverence for a man of such talent and such noble qualities of spirit. It was a searing pity for a man who had suffered so much without ever knowing joy and happiness, and who was so neglected by all his near ones...
>
> The dream of becoming his life's companion, of sharing his labours and lightening his existence, of giving him happiness – this was what took hold of my imagination; and Fyodor Mikhailovich became my god, my idol. And I, it seemed, was prepared to spend the rest of my life on my knees to him...

More often though, it was he who was on his knees to her, sometimes in person or, more frequently, on the page, making fevered promises to quit gambling,

> And here we are, I am yearning to see you, and you, nearly dying without me, my angel, I repeat, I am not reproaching you for that, and I love you even more for your yearning to see me. Having sent you the letters asking you to send me some money, I went to the gambling casino. I had altogether

twenty gulden left in my pocket (just in case), and I took a chance on *ten* gulden. I used almost supernatural efforts to remain calm and calculating *for a whole hour*, and wound up winning 300 gulden. I was so overjoyed, and had such a terrible, almost *mad* desire to finish off everything right there, *that very day*, to win perhaps twice that amount and then leave immediately, that I threw myself at the roulette table, without having given myself a chance to rest up and collect myself, took to betting gold pieces, and lost *everything, everything*, to the last kopeck, i.e., all I had left were *two* gulden for tobacco... I am leaving Hombourg with a loss. I also know that if I could give myself only four more days, I would surely win back everything. But, needless to say, I shall not play anymore!

But, of course,

Anya dear, my dearest, my wife, forgive me, don't call me a scoundrel! I am guilty of a criminal act, I lost everything that you sent me, everything, everything to the last kreutzer, I received it yesterday, and lost it yesterday. Anya, how am I going to face you now, what are you going to say about me after this!... Now I must hurry back to you. *Send me as soon as possible, this very moment, some money for my train fare – even if it is the last you've got.* ... Oh, if it hadn't been for that nasty, cold, and damp weather, at least I would have moved to Frankfurt yesterday! And nothing would have happened, I would not have played! But the weather was such that, with my teeth, and with my cough, there wasn't *a* chance to make a move, travelling the whole night in a light overcoat. It was simply impossible, it would have meant risking an illness. But now I shan't stop even for that. Immediately after receiving this letter send me...

ten imperials, i.e., a little over 90 guldens, so that I can pay my bills and my train fare...

Reunited, their life continued in this vein. They travelled to Baden-Baden together, the original of Roulettenburg.

FM used to return from roulette pale, exhausted, barely keeping on his feet, would ask me for money (he entrusted all money to me), leave, and after half an hour, return still more disconcerted, for money, and this up until the time he had lost everything we had with us. When there was nothing with which to go to the tables and there was nowhere to obtain money, FM used sometimes to be so overwhelmed that he began to weep, got down on his knees before me, imploring me to forgive him for tormenting me with his behaviour, would go into extreme despair.

In Baden, they stayed originally at the Hotel Zum Goldenen Ritter, but then, short of money, they moved to above a blacksmith's shop, where work began at four in the morning.

I have no wish to do anything, either to sew or write or read (it is true, there is positively nothing to read). Fedya advises me to go and read at the library which is attached to the casino; he says that women go there... I should just like to lie in bed with the curtains closed and to think of nothing... at last Fedya came. At first he said, 'no luck', and then showed me a full purse. It contained sixty-one louis, and with our thirty it makes ninety-one. ... Fedya told me that he had had a run of luck and everybody wondered at it: he won on whatever he staked. Behind him stood an Englishman [who] staked on the same *numbers* as Fedya did; and Fedya observed that each time he made a stake and looked at the Englishman, he won without fail: such a lucky face that man has. Fedya says that the Englishman's face is so good and kind that it is bound to bring luck...

I went for a very long walk, and coming back I met Fedya, who said that he had lost all and was waiting for me. He said he had lost because behind him stood a rich Pole and a young little Pole, who made very small stakes, but gave themselves airs. This made Fedya so angry that he played carelessly and lost.

Luck came with an Englishman's face (Mr Astley in *The Gambler* is the embodiment of prudence); it departed with the arrival of two Poles.

Looking back at that time, four years later, Anna wrote,

Fyodor so often spoke of the certain 'ruin' of his talent, if we remained any longer abroad, and was tormented by the thought that he would not be able to keep his family, that, as I listened to him, I too was driven to despair. To relieve his anxiety and disperse his gloomy thoughts, which prevented him from concentrating on his work, I had recourse to the device which always helped to distract and amuse him. As we possessed then about three hundred thalers, I said that it would be worthwhile to try once more our luck at roulette. I pointed out that as he had occasionally happened to win, there was no reason why we should not hope that our luck would turn this time. I certainly did not entertain any hope of his winning at roulette, and I also was very sorry to part with the three hundred thalers, which it was necessary to sacrifice; but I knew by experience of his former visits to the tables that, after those exciting emotions, after satisfying his craving for risk, his passion for gambling, Fyodor would return home calmed, and that then, realising the futility of his hopes of winning at the tables, he would sit down with renewed strength to his novel, and make up for all the lost time in a couple of weeks.

In April 1871, Dostoevsky followed Anna's advice and returned to Wiesbaden.

Wiesbaden 28 April

Anya, for the sake of Christ, for the sake of Lyuba [their child], for the sake of our whole future, don't start worrying and getting all upset – read this letter carefully to the end. You will see at the end that disaster isn't really a reason for despair, but on the contrary, something may even have been gained by it which will be much more valuable than the price paid for it! And so calm yourself, my angel, hear me out – read this to the end. For Christ's sake, don't fall to pieces.

You, my precious one, my lifelong friend, my heavenly angel, you have, of course, gathered that I have lost everything – the whole of the 30 thalers you sent me. Remember that you are my only salvation and that there is no one else in the world who loves me. Remember, too, Anya, that there are misfortunes that carry their own punishment.

Now Anya, you may believe me or not, but I swear to you that I had no intention of gambling!... When I got the 30 thalers today, *I did not want to gamble* for two reasons: (1) I was so struck by your letter and imagined the effect it would have on you (and I am imagining it now!) and (2) I dreamed last night of *my father* and he appeared to me in such a terrifying guise, such as he has only appeared to me twice before in my life, both times prophesying a dreadful disaster, and on both occasions the dream came true. (And now, when I think of the dream I had three nights ago, when I saw your hair turn white, my heart stops beating – ah, my God, what will become of you when you get this letter!)

But when I arrived at the casino, I went to a table and stood there placing imaginary bets just to see whether I could guess right. And you know what, Anya? I was right about ten times in a row, and I even guessed right about Zero. I was so amazed by this that I started gambling and in 5 minutes won 18 thalers. And then, Anya, I got all excited and thought to myself that I would leave with the last train, spend the night in Frankfurt, and then at least I would bring some money home with me! You had pawned all your possessions for me during these past 4 years and followed me in my wanderings with homesickness in your heart! Anya, Anya, bear in mind, too, that I am not a scoundrel but only a man with a passion for gambling.

(But here is *something else* that I want you to remember, Anya: I am through with that fancy forever. I know I have written you before that it was over and done with, but I never felt the way I feel now as I write this. Now I am rid of this delusion and I would bless God that things have turned out as disastrously as they have if I weren't so terribly worried about you at this moment. Anya, if you are angry with me, just think of how much I've had to suffer in the coming three or four days! If, sometime later on in life you find me being ungrateful and unfair towards you – just show me this letter!)

By half past nine I had lost everything and I fled like a madman. I felt so miserable that I rushed to see the priest. As I was running toward his house in the darkness through unfamiliar streets, I was thinking: 'Why, he is the Lord's shepherd and I will speak to him not as to a private person but as one does at confession.' But I lost my way in this town and when I reached a church, which I took for a Russian church, they told me in a store that it was not Russian but yid. It was if someone had poured cold water over me. I ran back home. And now it is midnight and I am sitting and writing to you.

The letter goes on, Dostoevsky's usual sort of fussing over the details of arrangements to be made for the transfer of money so he can travel back to her. And it's in his customary heightened language, the abasing writer, the manipulated reader pushed between opposing feelings and thoughts, the voice of hyper-self-consciousness, of mind and spirit at the very edge of itself. As Dostoevsky wrote of himself, 'The worst thing is that my nature is base and too passionate. Everywhere and in everything I go to the limit. All my life I have crossed the last line.'

But, something has changed, even if it all seems to be the same,

Anya, I prostrate myself before you and kiss your feet. I realise that you have every right to despise me and to think: 'He will gamble again.' By what, then, can I swear to you that *I shall not*, when I already have deceived you before? But, my angel, I know that you would die (!) if I lost again! I am not completely insane, after all! Why, I know that, if that happened, it would be the end of me as well. I won't, I won't, I won't, and *I shall come straight home!* Believe me. Trust me for this *last time* and you won't regret it. Mark my words, from now on, for the rest of my life, I will work for you and Lyubochka without sparing my health, and *I shall reach my goal!* I shall see to it that you two are well provided for.

You mustn't think I'm crazy, Anya, my guardian angel! A great thing has happened to me: I have rid myself of the abominable delusion that has *tormented* me for almost ten years. For ten years (or, to be more precise, ever since my brother's death, when I suddenly found myself weighted down by debts) I dreamed about winning money. I dreamed of it seriously, passionately. But now it is all over! This was the *very* last time. Do you believe now, Anya, that my hands are untied? – I was tied up by gambling but now I will put my mind to worthwhile things instead of spending whole nights dreaming about gambling, as I used to do. And so my *work* will be better and more profitable, with God's blessing! Let me keep your heart, Anya, do not come to hate me, do not stop loving me. Now that I have become a new man, let us pursue our path together and I shall see to it that you are happy!

It continues, more recriminations, the usual Dostoevskian identification of command of self with being lucky, or finding the seam of luck, another reminder about sending the money and the travel arrangements, again the promise not to see the priest, multiple exclamation marks, multiple postscripts and the third PS,

I will remember this as long as I live and each time I think of it I will bless you, my angel! Let there be no mistake, now I am yours, all yours, undividedly yours. Whereas, up till now, *one half* of me *belonged* to that accursed delusion.

His wife was sceptical,

I, of course, couldn't all at once believe in such great happiness as Fyodor's indifference to roulette. He had promised me not to play so many times before, and never found the strength to keep his word. But this time the happiness was realised. That was indeed the *last* time he played roulette. ... He returned from Wiesbaden cheerful and calm, and immediately sat down to the continuation of his novel, *The Devils*.

Anna was pregnant again and maybe Fyodor at last was recognising a greater imperative than his success, and his dream of paying off all his debts and obligations. Maybe God was intervening. There is no doubt that he was sincere, and he didn't gamble again – but there were deceits even here: despite what Dostoevsky says, his gambling long predated Mikhail's death and his own assumption of his brother's debts.

That zero Fyodor mentions in his letter to Anna is a reference to Aunty in *The Gambler*, the character's compulsion to keep making the ruinous, luck-defying bet, and a sentimental reference to the couple's early time together, their shared project of his novel that had saved him before from his own ruin. Joseph Frank, Dostoevsky's pre-eminent biographer, says, 'It is all the more remarkable that he broke the spell in 1871, just before returning to Russia, and recorded this resolution in a letter describing how he entered a Jewish synagogue by mistake while roaming through the darkness of unfamiliar German streets in search of a Russian church. Dostoevsky believed in omens, and there is probably a close connection between this seemingly insignificant incident and the definitive termination of his gambling mania.'

The synagogue connotes those 'inhuman' Jews who could, unlike Dostoevsky, master their own will at the gaming table.

The dreams of his father would have had a significance too, for me perhaps more than for Dostoevsky.

I'm on my way to the casino in Baden-Baden, to put the *Gambler* 'system' into operation, and I've been rereading *The Gambler* and reading Dostoevsky's letters and his wife's diaries – and this talk of Jews, and fathers, brings me inevitably to think about, and dream about, my own father, who shared the writer's animosity towards Germans and could be just as rude about Poles, although my father's rudeness would extend to Russians also.

✳

When I went to the Grand Casino in Baden-Baden, it was with a very cool head. I had to rent a tie at the reception desk in order to conform to the casino's dress code. But I remained undeterred, even if I was in 'unlucky' clothes: my black moleskin suit in which I have seldom won at poker.

If Dostoevsky were gambling now, he would be playing poker rather than roulette. Roulette does not reward composure: it is the operation purely of chance. Poker does reward composure, and nerve, and luck.

But I was on his territory, his world, here to play his game. Affecting nonchalance beneath muralled ceilings, high mirrors, chandeliers, I went across the red carpet to a roulette table where I tried to put the Dostoevsky betting 'system' into operation. Twice, as Aleksy had done in *The Gambler*, I bet on *passe*. Each time I lost. I put a €10 chip on red. The croupier spun the wheel, an elderly woman, bare-armed in a blue evening gown, placed her counters on, it seemed, half the numbers on the board until she was told in French that it was too late to make any more bets. The wheel spun, the ball skimmed and bounced and landed in a black slot. I had lost again. So had the elderly woman, as the croupier placed a metal figure on the winning number and shovelled all her counters off the table.

The Dostoevsky 'system' was resisting me. I couldn't get past step one. Nor could I establish any complicity with my fellow gambler, who ignored my friendly look, my attempt to console her, and myself, on our bad luck, as she laid out her counters for the next coup.

I moved to the poker table, where I started nervously. But within an orbit of a game (the dealer button moves in a clockwise position with every deal; when everyone has 'been' the dealer, an orbit is complete), it becomes clear who the good players are, or, in this case, is. Most games revolve around providers and tourists and will have one or two regular sharks who feed on them. The shark in Baden-Baden was a tanned middle-aged Italian in a black suit of a more tailored cut than mine and crisply gelled hair. In our time at the table together I heard him speak at least four languages, all with some fluency, and he made for pleasant, if utterly guarded, company. Prudently, I changed seats so I was sitting to his left rather than to his right, which gave me 'position' over him, meaning that he had to make his actions before I made mine. I settled down, committed one mistake, not to call the bluff of a burly Romanian, but otherwise, I played optimal poker.

The Italian, in German, asked after the family of the dealer. The regular has relationships with the cardroom staff and management. It is in his interests to make his, and their, working lives as pleasant, smooth, congenial and informative as possible. There is no question of impropriety here, on either side; and when he gives little presents at Christmas, this is not in the nature of a tip or an incentive or a thank you for services rendered, it's a simple holiday gift from one friend to another. They are almost colleagues.

I played well. I kept out of the Italian shark's way and we took it in turn to build up our stacks against the tourists.

And I was still €300 down on the night. All because of one hand, in which I'd flopped an up and down straight flush draw, four-bet my opponent, who happened to be holding the nuts on the flop and wasn't getting away from anything. My hand didn't improve and I was left on the wrong side of an €800 pot.

But what was I doing here? I was losing money in the same casino as Dostoevsky, but hardly experiencing the passions and humiliations that went into the writing of *The Gambler*. This is the town where Turgenev had lent him money that both knew he was never going to repay. This is the casino that he returned to, while his new wife waited for him in the apartment above the blacksmith's shop, confiding her discontents to her diary.

I had hoped that this was going to be my point of entry to the great, sickly man, of whom Nietzsche had written that he was the only psychologist from whom he had anything to learn, that encountering Dostoevsky's work 'was the greatest accident of my life'. The following morning I took a walk through the park that runs from the back of the hotel, which was where Turgenev had his house. By being here I had learned that the blacksmith's shop was now an estate agent's, that gentlemen have to wear ties before they may lose their money at the Grand Casino, and I felt further away from Dostoevsky than ever.

A few days before coming out here, I'd renewed my relationship with poker at the Grosvenor Victoria Casino on the Edgware Road, the 'Vic'. After putting my name on the waiting list, I'd gone down to the casino floor to watch the roulette tables. Sitting in an armchair I fell into conversation with the gambler sitting in the chair beside mine.

'That's nice shoes,' the older gentleman said, with an accent that I placed as Greek.

'Thank you,' I said.

'Must be sixty-pound shoes. How much?'

'How much did they cost? I'm not sure, they were a gift.'

I knew perfectly well how much they had cost, and it was more than sixty pounds.

'Look at my shoes. Twenty-pound shoes. My shoes are shit. Isn't it.'

His shoes were brown and plastic and too big for his feet. My name was called on the public address to let me know that a seat was free on the £1–2, but I didn't know how to end the conversation. He told me that he lived on his own, that his son kept an eye on him, made sure he ate enough, and that this being Tuesday there was still some money left in his weekly pension to pay for his roulette habit. Which seemed to be the cue for us both to take our places at our respective tables.

I might have gained more insight into Dostoevsky by talking further with the old gambler. But instead I was in Baden-Baden, where my destiny was being untouched, where nothing radical or definitive seemed to be happening to me.

It was in Wiesbaden, not here, that Dostoevsky made his finally successful resolution to quit gambling. And it was at the casino in Wiesbaden that the ruined young gambler had shot his brains out, giving Dostoevsky the first stimulus to write *The Gambler*. I could have gone there, lost money at the casino in Wiesbaden, tried to relive the dreamlike walk Dostoevsky described to his wife, looking for a Russian church and finding a Jewish synagogue instead. I had come rather to Baden-Baden because that is supposedly the original of the town of Roulettenburg.

But 'Roulettenburg' isn't Baden-Baden or Wiesbaden or Bad Hombourg, it's a mixture of the three, and none of them. It's a fictional space, which means it exists only on the page, and in the minds of writer and reader. Going to Wiesbaden instead of Baden-Baden wouldn't have brought me closer or pushed me further away from Dostoevsky. Maybe I should have gone to Siberia instead.

Dostoevsky had been sentenced to death in 1849, for 'plotting to subvert public order'. After enduring a mock execution in St Petersburg, the three 'conspirators' learned that their sentence had been commuted to four years hard labour in Siberia, followed by five years in military service. Dostoevsky's memoir of penal servitude, *In the House of the Dead*, was Wittgenstein's favourite of his books.

My father too was sentenced to hard labour in Siberia. He and his friend Benny Zysmanowicz, teenage communists, had left Nazi-occupied Warsaw for the Soviet Union. In November 1939 they made it across the Biała river with coins sewn inside the lining of their overcoats, and quickly became part of the great mass of displaced Poles on the newly Soviet-occupied side of the border, in what is now Belarus.

As well as the coins, Izio, as my father was named then, and Benny had about two hundred razor blades. 'So we started speculating, selling razor blades... We got space in a kitchen in someone's house until the money ran out. Before it started running out actually, we went to volunteer for work in the Ural mountains. We came in front of the interviewer and said we want to be together.' But they weren't allowed to be together. Izio was sent to Kamensk, to work in a magnesium factory. Benny came to visit him once and they agreed to meet again in Moscow for the May Day parade, but they never saw each other again.

Izio Flusfeder had a talent for making friends at that time. In Kamensk, he became friends with an editor of *Der Moment*, a Warsaw Yiddish newspaper. Fifty-four years later, when I was questioning him about this part of his life, and he was consenting to answer my questions with some degree of patience, he couldn't remember the editor's name. Together they went to Moscow, where he had hoped to see Benny again. 'I was arrested four times in two days, stuck out like a sore thumb. We got back to the border, not Białystock but Kovel.' The editor found work as a nightwatchman. Izio got a job as a cowboy. Three days later, he waited in line along with thousands of others to register with the Germans to get back home. But he was told, 'We don't register Jews, we only register gentiles'.

He lost the cowboy job, found another carrying cement on a construction site, soon collapsed, went to hospital for ten days, and after he was released, he heard that the Russians were also registering Polish nationals. No news was reaching them from Warsaw, but he had fallen out of faith with the workers' paradise and wanted to go home. 'Again, thousands and thousands of people stand in line. Within a few days we were all arrested. After being held for a few days between a school and prison, they shipped us in cattle cars, they were telling us we were going back to Poland, two weeks later we were in Siberia. The train ride was a nightmare. No food. Some days they'd bring in water. Some days nothing. There was no planning. There was no mood. There was just depression. That was the end. You just obeyed. Whatever they told you. There was a hole cut out in the floor. Women, men, whatever, would use it.'

These quotes from my father are transcribed from a tape recording I made asking questions about his past in New York City in 1993. My wife and I were visiting from London.

I have the tapes still, my father's measured deep voice, his accent a beautiful mix of Polish and English and American, my lighter voice prompting him.

I think of him as Joe Flusfeder, but back then, at the end of the railway line sleeping outdoors in a Siberian winter, he was still Izio, wearing the remnants of his grey double-breasted suit from Warsaw, the one he had been wearing when he had gone to Red Square, in the company of the editor from Warsaw whose name he could no longer remember, hoping to see Benny Zysmanowicz again.

'At that time they started dying like flies. From the cold, from dysentery.'

By January they were in underground shelters, with stoves to keep them warm. 'There were only six hundred of us that survived.' There is maybe some quiet pride in the way he says this, his estimation that not much more than a quarter of his cohort made it, and I begin to push him on what might have enabled this exceptionalism, and what was in his thoughts.

So you expected to die?

And very quickly the note of pride is lost, in some impatience with the impossibility of me, in my soft existence, being able to comprehend his experience. 'You don't think about those things. Your only wish is to fill up your belly. You don't philosophise. You can't. I don't believe you can understand what's going on in a mind like that.'

Unlike Dostoevsky, my father did not have to undergo a mock execution, but like Dostoevsky, he ended up a soldier, but not for the Russians.

Do you remember what you were thinking at the time?

'How to survive. That was about it… See what happens, see what happens… At that point my only need was to get back to my family that were still sitting there waiting for me. I'm going to Warsaw because that's where they are. If they are on the moon that's where I'm going because that's where they are. In the camps it started. I saw the steps leading up to the apartment. I saw the nameplate on the door. Brass, engraved, very flowery sort of script. We were on the first floor.'

Meanwhile, his family were dying. One aunt and one uncle survived. His brother, David, was shot by a German soldier in the street. His father, Szlama, was transported to the camp at Trawniki, where he died. All his cousins and most of his aunts and uncles were killed in the death camps. His mother, Helena, who had breast cancer, committed suicide in the Warsaw Ghetto after the deaths of her husband and older son, and the presumed death of her younger one.

What did it do to you, finding this out?

'The actual experience probably dehumanises you personally. But when you look at the overall picture, you tend to think we live in a jungle, there's no purpose, no reason for it, and somebody starts talking to you about God, how are you going to react to it? Good, bad, or indifferent, if he's that bad, who the fuck needs him? It's possible that what you saw you saw, your own experience, the cruelty of man to man, and that goes on unpunished, what are we doing... going back to pagan times and creating a golden calf and praying to it, that's basically what we're doing.'

Neither Dostoevsky nor my father came out of Siberia with their socialist views intact. My father never had believed in God; his faith in a progressive collective was replaced by a sense of his own capacities to survive, and some guilt that he had done so. But I'm getting this slightly wrong: he did have a belief, never mystical or sentimental, in luck.

When Joe Flusfeder talked of Izio's experiences, in Russia, Iran, or in Italy, at the Battle of Monte Cassino, where he was one of the few Jews in the battalion and was ironically nick-named Jacek, he was a seventy-one-year-old remembering a younger, differently named man. And from this distance, he refers to much of his earlier experiences and thoughts in a further distancing second person.

You must have thought you didn't deserve to get through this…

'Well, that's one of the things probably. That's one of the major things, if there's any kind of righteousness in this world… But there is no righteousness, there is just a matter of simple coincidences, of a series of coincidences. You always come to a fork in the road, take a left it's fine, take a right, you die. So you have to keep on taking the left… What are your chances? You really don't know, if you'd taken a different road, whether things would really have been different. I tried not to think about it… Sometimes it comes back and you deal with it; that's what happened and you try to make the best of it.'

After he retired, from a career as a mechanical engineer, on the profits from the automated record press he had co-invented, my father played patience almost endlessly. He seemed to find a comfort in submitting to a closed structure over which he had little control. Occasionally, on vacation, he would play blackjack at a casino, but generally he didn't choose to engage in any activity where the odds were against him. He disapproved of gambling but not in a moral way. He said that when a bet was made, there were two participants, a fool and a thief. Or, as he said it, because his Polish tongue was never able to produce the English *th* sound, *a fool and a t'ief*. He preferred to take his chances in the world, against other men.

*

After Wiesbaden, and his renunciation of gambling, the Dostoevskys returned to Russia, where gambling was illegal, and even though they made four more European trips in the 1870s, 'he never once considered going to a gambling resort', Anna wrote. 'It seems that his fantasy of winning at roulette was a kind of obsession or disease from which he recovered suddenly and forever.'

Although, maybe all Dostoevsky had done was to accept that he would never be able to retain his composure when facing a roulette wheel. In his late novel *A Raw Youth*, published in 1875, Dostoevsky, by now the revered great writer, the representative of Russian soul and nativism and Orthodox Christianity, wrote, barely disguised as his character Arkady, 'I still retain the conviction, that in games of chance, if one has perfect control of one's will, so that the subtlety of one's intelligence and one's power of calculation are preserved, one cannot fail to overcome the brutality of blind chance and to win.'

roulette –£30
poker –£260

CHAPTER 5

Getting God to Speak

Lots are cast into the lap, but they are disposed
of by the Lord.

Proverbs, 16:33

net total −£300

My neighbour Barbara showed me the lottery ticket she had
just bought.

'I don't usually play it,' she said.

I asked her what had changed.

'The jackpot! It's rolled over to sixteen million!'

Barbara said that people tend to use significant dates like
birthdays, which means that numbers cluster below 32,
dividing the biggest prizes among multiple winners. She had
chosen high numbers, because she didn't intend to share her
jackpot. She was sure that she was going to win it. She had,
she told me, a feeling.

'Someone has to win it,' she said.

This is true. People do win on the lottery. Voltaire made
his fortune on it. So did Casanova. And so did the mysterious Joan Ginther, once of Bishop, Texas, now resident in Las
Vegas, Nevada.

'You should understand,' Barbara said. 'You're a gambler,
aren't you?'

'Well, poker. Which isn't quite the same thing.'

She wasn't really listening. Barbara had bought three tickets, which meant her odds of winning were more than five million to one.

'It's meant to be,' she said.

*

There are few accidents in classical texts; the gods are behind it all. It is not a storm at sea that Homer's Odysseus suffers, but Poseidon's rage. In Virgil's *Aeneid*, Juno may not influence the weather directly, but works her malice through lesser gods. Her proxy, Aeolus, the god of the winds, causes the Trojan shipwrecks and blows Aeneas off course. The subsequent turns in the hero's fortunes are impelled by divine intention, whether it be the benevolence of his mother Venus or the continuing malice of Juno. There's not much luck here.

Except in the case of Dido. The Queen of Carthage is repeatedly described by Virgil as *infelix*, which, like the German *unglücklich*, means both 'unhappy' and 'unlucky'. She loses Aeneas, not because of any lack or inequality in their love for each other, but because he must fulfil his destiny, the divine command to found Rome. Virgil uses a rare simile of accident to describe her intense love,

Unlucky Dido burns, and wanders through the city in a frenzy – even as a doe in the Cretan woods, unwary, smitten by an arrow, which an unknowing shepherd hunting with darts has pierced from afar, leaving in her the winged steel: she zig-zags through the Dictaean woods and glades, but fast to her side clings the deadly shaft.

An 'unwary' deer, an 'unknowing' hunter far away; neither of them has seen the other, or has any knowledge of the connection between them; but the arrow has landed, the deer is dying, and unlucky, unhappy Dido is pierced, fatally. She falls on her own sword after Aeneas sails away to follow his divine path.

Dido, the bad luck casualty of Aeneas's great good fortune, exemplifies the ancients' view of the nature of love – a delirium of the senses that might cause empires to fall, that can overturn anything. Luck's martyr has to die. Their love must not be allowed to thwart the hero's destiny.

If there are few accidents in the classical texts, if the gods, or Fortune, or God, make all things happen, if events have a divine cause, and interpretation, then not only is it reasonable to suppose that things are 'meant to be' – every lottery win, and loss, has already been decreed – it must also be reasonable to suppose that we might pray for an intercession in our affairs, and use oracles to uncover the plan, the secret hidden pattern, and to ask for advice.

Aeneas knows where to found Rome because 'every sign from heaven – the stars, and tongues of birds and omens of the flying wing – has uttered favourable words to me about my journey, and all the gods in their oracles have counselled to make for Italy'. In the earliest works of literature, characters seek and find prognostications to uncover their fate.

It is but a small jump for readers to do the same. In Wilkie Collins's *The Moonstone*, the venerable Gabriel Betteredge turns to random pages of Daniel Defoe's *Robinson Crusoe* whenever he requires assistance or guidance,

I am not superstitious; I have read a heap of books in my time; I am a scholar in my own way. Though turned seventy, I possess an active memory, and legs to correspond.

You are not to take it, if you please, as the saying of an ignorant man, when I express my opinion that such a book as *Robinson Crusoe* never was written, and never will be written again. I have tried that book for years – generally in combination with a pipe of tobacco – and I have found it my friend in need in all the necessities of this mortal life. When my spirits are bad – *Robinson Crusoe*. When I want advice – *Robinson Crusoe*. In past times when my wife plagued me; in present times when I have had a drop too much – *Robinson Crusoe*. I have worn out six stout *Robinson Crusoes* with hard work in my service.

Gabriel, as many of his readers knew, was following the example of his hero. Crusoe also combines the pleasures of tobacco with the practice of bibliomancy,

I took up the Bible and began to read, but my Head was too much disturb'd with the Tobacco to bear reading, at least that Time; only having opened the Book casually, the first Words that occurr'd to me were these, 'Call on me in the Day of Trouble, and I will deliver, and thou shalt glorify me.'

The tobacco is relatively new, but the bibliomancy is part of a long tradition. In his *Confessions*, his exemplary memoir of a soul moving from darkness into light, the church father Augustine recalls a child's voice instructing him to, 'Take up and read; take up and read.' Opening the Bible to 'the first chapter I should discover' ('casually', as Robinson Crusoe would say – or randomly, as we would) he reads a passage from the book of Romans that exhorts him to 'walk honestly as in the day: not in rioting or drunkenness, not in chambering and impurities, not in contention and envy. But put ye on the

Lord Jesus Christ; and make not provision for the flesh in its concupiscences.' The passage seems to speak to him directly, 'No further would I read, nor needed I; for instantly at the end of this sentence, by a light as it were of serenity infused into my heart, all the darkness of doubt vanished away.'

Later he turned against the practice, 'As to those who read futurity by taking a text from the pages of the Gospels, it is better that they should do this than go to consult spirits of divination; nevertheless I am displeased with this custom, which turns the divine oracles, which were intended to teach us concerning the higher life, to the business of the world and the vanities of the present life.'

It's not that it's ineffective to use the sacred books in this way. Augustine seems to have no doubt that we can read futurity in a random Bible text, casually chosen, 'discovered', but his disapproval rises against using something holy for frivolous purposes – and contained within it is a revulsion against popular practices of the recent pre-Christian past.

The *sortes Biblicae* or *sortes Sanctorum* that Augustine is so disapproving of follows the principle of *sortes Homericae* or *sortes Virgilianae*, originally a form of soothsaying in which a random slip was chosen on which were written verses by Homer or Virgil. This was later simplified by opening a copy of the *Iliad* or the *Aeneid* itself and interpreting as prophetic the first line that the eye fell upon.

The fourth-century *Historia Augusta* gives many examples of Roman emperors consulting Virgil to uncover their own destinies. Gordian II opened the *Aeneid* at book 1, line 278, 'For these I set no bounds in space or time; but have given empire without end'. Less auspiciously, the *Aeneid* predicted the short reigns of Claudius II and Quintillus. Hadrian drew book 6, line 808 ('But who is he apart, crowned with sprays of olive, offering sacrifice? Ah, I recognise the hoary hair and beard of that king of Rome who will make the infant city secure on a basis of laws') which was taken as predicting his adoption by Trajan and succession to the imperial throne. Alexander Severus turned to book 6, line 851 ('You, Roman, be sure to rule the world').

In the period of unrest between parliament and king that would culminate in the English Civil War, Lucius Carey once went to a public library in Oxford with Charles I and, being shown a finely printed and bound copy of the *Aeneid*, suggested to the king that he use the *sortes Virgilianae* to tell his future. The king opened the book and happened on Dido's prayer against Aeneas in book 4, 'Nor yet, when he has submitted to the terms of an unjust peace, may he enjoy his kingdom or a longed-for light, but die before his time and lie unburied on the sand.' This did not feel like good news. To cheer him up, Carey chose his own passage, and his too was an episode of untimely death, which his friends later took as presaging his own a couple of years later at the Battle of Newbury in 1643.

During his final illness, Dostoevsky instructed his wife, Anna, as she had done before, to open the New Testament at random and read the ensuing passage. She happened upon the account in Matthew 3:15 of Jesus coming to John to be baptised. 'And Jesus answering, said to him, Suffer it be so now. For so it becomes us to fulfil all justice.' Dostoevsky said, 'Do you hear? "Suffer it to be so now." Don't hold me back. Of course I am about to die.'

But why Virgil rather than Ovid? Why *Robinson Crusoe* rather than *Moll Flanders*? Why the New Testament rather than the Old? Why the *I Ching*? Why should any particular book have more power, or access to hidden truths, than any other? Jewish worshippers touch a kiss to a prayer book when it is picked up and laid down, they apologise to it should it fall. As if God inheres more deeply in some objects, and places, and procedures, than others.

*

The foundational event of the Judeo-Christian traditions is the story of Moses on Mount Sinai. According to the book of Exodus, Moses was on the mountain for forty days and nights, culminating in God giving him the tablets of stone on which were engraved the Ten Commandments, so establishing an ethical tradition and a set of communal obligations.

There were also extensive instructions about how the priesthood should operate. These would have been added later, after the priests had established their power, to give their cultic operations scriptural authority.

Within the priestly instructions is an account of the mysterious Urim and Thummim, which were to be placed in the priest's 'breastplate of judgement'. These seem to have been stones, or jewels, that stood for yes and no, so that questions could be asked, about the guilt of accused criminals, the distribution of property, directions for correct behaviour, and the answer, God's word, would be derived from how the stones fell.

(The legend goes that God also told Moses something else, which wasn't written down, an oral revelation to go along with the inscribed one; and the work of mystics ever since has been to get at this spoken, other truth.)

As in the classical texts, there are few accidents in the Bible. Events, even the most seemingly chanceful, are manifestations of the divine will.

The story of the first king of Israel begins with the shepherd Saul straying far from home in search of his father's lost asses, and finding his way to the prophet Samuel, who anoints him as the divinely ordained king he has been waiting for. The truth of Saul's divine destiny is confirmed when Samuel draws lots. Throughout his reign, Saul will ask God questions, until 'the Lord did not answer him, neither by dreams nor by the Urim nor by prophets' (1 Samuel 28:6). David on his way to supplanting Saul as king often 'enquires' of God. These enquiries are binary: he asks a question and God replies with a yes or a no. In other words, as Saul was, and Samuel had, he is consulting the Urim and Thummim, or drawing lots.

The English word 'lot' is of German origin, from *hleut*, which designated the bean or pebble or token or knuckle bone or die – or the stones or gems of the Urim and Thummim. It is a means of getting God to speak.

This is true in the New Testament as well as the Old: Matthias was chosen to replace Judas as the twelfth disciple by casting lots; Jesus's clothes were distributed by lot after his crucifixion. The procedure in both is the same, but the idea is somehow different: there's no suggestion that God is choosing who gets to wear what; but the appointment of Matthias is made according to God's will.

Theological justification for this was provided by the seventeenth-century clergyman Thomas Gataker, who made the distinction between 'ordinary' and 'extraordinary' lots. Gataker argued that when there is little chance of one particular outcome – he gives the example of drawing lots to determine which minister of a hundred should visit a plague house – and it is therefore evident that it would most likely have a different outcome, how then are we to claim that a single draw is evidence of God's will? The inference from this stochastic reasoning is that the more probable something is to come to pass, then the more certain we may be that it is in line with God's wishes. God, Gataker seems to say, backs the favourite. Unpredictability is unsacred. And certainty is self-evidently morally right.

This separation of 'casual events' from revelations of the divine will makes it permissible for gamblers to speculate on them, and for the exchequer to profit on the speculations. Public lotteries are lawful if 'used with due Caution… wherein men buy bare hope alone then [than] actually ought else'.

*

Augustine had been one of many early Christians to denounce augury by bibliomancy. The frequency and force of the denunciations show how popular the practice must have been. 'From the fourth to the fourteenth centuries, these *sortes Sanctorum*,' Edward Gibbon wrote, 'were repeatedly condemned by the decrees of councils and repeatedly practised by kings, bishops and saints.'

These condemnations are echoes of the proscriptions made in the Bible against any kind of fortune-telling. Leviticus 19:26 commands us not to 'practise divination nor to observe times'. The rabbinical sage Rashi glosses this as meaning not 'to draw prognostications from the cry of a weasel or the twittering of birds, or from the fact that the bread falls from his mouth or that a stag crosses his path', or that 'this or that day is auspicious for beginning a work, this or that hour is inauspicious for starting on a journey'.

As ever, the Bible gives mixed messages. Paul's epistle to the Thessalonians instructs them not to despise prophesying. The prophet Elisha in Kings II 13:15 uses the shooting of arrows as a divining tool to prophesy the victory of Joash over the Syrians. Joseph gains power in Egypt by interpreting the Pharaoh's dreams and in Genesis 44:5 there's a reference to the goblet he uses for divination, which presumably is a hydromancy similar to the ancient Greek practice of asking a question before tossing gold or jewels into a vessel and interpreting the ensuing ripples. As with consulting the *Aeneid*, a paradoxical principle is at work here, from which most augury and mysticism derive, that a random procedure gives access to a fixed truth. This notion, of how we get God to speak, is as close to the centre of the Judeo-Christian traditions as it is those of Greece and Rome.

Praying to God for guidance, for one's own mind or conscience to be alerted to the correct path to follow, or opening the Bible at random, or drawing lots – these are all congenial to Protestantism: the visible world reveals the hidden truth to the individual believer, without the intercession of priests or external authority.

The Methodist John Wesley used the drawing of lots to decide whether to marry or not. Wesley arrived in Georgia in March 1736, twenty-eight years old, a missionary pastor. He wrote in his journal, 'At my first coming to Savannah… I was determined to have no intimacy with any woman in America.' Before the month was out, he was already feeling tested. He wrote to his brother Charles – in Greek, to keep the letter confidential, away from eyes, evil or otherwise, 'I stand in jeopardy every hour. Two or three are women, younger, refined, God-fearing. Pray that I know none of them after the flesh.'

The prime temptation was Sophy Hopkey, the seventeen-year-old niece of the chief magistrate of the colony. Her aunt and uncle were delighted in the growing friendship of the two. By October, her uncle told Wesley, 'Do what you will with her. Take her into your own hands.' Mr Wesley became Miss Hopkey's instructor in French and prayer.

My desire and design still was to live single... My greatest difficulty was while I was teaching her French, when being obliged (as having but one book) to sit close to her, unless I prayed without ceasing I could not avoid using some familiarity or other which was not needful. Sometimes I put my arm round her waist, sometimes took her by the hand, and sometimes kissed her.

He thought of marrying her. Others advised him to do so.

I find, Miss Sophy, I can't take fire into my bosom, and not be burnt. I am therefore retiring for a while, to desire the direction of God. Join with me, my friend, in fervent prayer, that he would show me what is best to be done.

Finally, he asked God to speak directly. In the company of his friend and travelling companion Charles Delamotte, Wesley made his appeal to the 'Searcher of Hearts'. Here is part of his journal entry from 4 March 1737,

Having both of us sought God by deep consideration, fasting and prayer, in the afternoon we conferred together but could not come to any decision. We both apprehended Mr Ingham's objection to be the strongest, the doubt whether she was what she appeared. But this doubt was too hard for us to solve. At length we agreed to appeal to the Searcher of Hearts. I accordingly made three lots. In one was writ, 'Marry': in the second 'Think not of it this year'. After we had prayed to God to 'give a perfect lot', Mr Delamotte drew the third, in which were the words, 'Think of it no more'. Instead of the agony I had reason to expect I was enabled to say cheerfully 'Thy will be done'. We cast lots again to know whether I ought to converse with her any more, and the direction I received from God was 'Only in the presence of Mr Delamotte'.

On 8 March, Sophy breakfasted with Wesley and Mr Delamotte. Wesley asked her if it was true that her uncle's clerk, William Williamson, 'pays his address to you'. Sophy replied that she had 'no inclination' for Williamson. But the next morning her aunt asked Wesley to publish the banns of marriage between Sophy and Mr Williamson.

Wesley's flirtation with Miss Hopkey had been interfering with his plan of performing missionary work to the Indians. He was still the preferred groom, both for Sophy and her family, but he refused to make the proposal. It seemed that his vocation and marriage were mutually exclusive.

To see her no more! That thought was the piercings of a sword. It was not to be borne – nor shaken off. I was weary of the world, of light, of life. Yet one way remained: to seek God, a very present help in time of trouble. And I did seek after God, but I found him not. I forsook him before; now he forsook me. I could not pray. Then indeed the snares of death were about me; the pains of hell overtook me.

On the day of Sophy's marriage to Williamson, Wesley wrote his will.

On 11 July, he noted in his diary, 'Mrs Williamson miscarried.'

Wesley was forced to leave Savannah in the aftermath of the court case against him after he refused to give the sacrament to Mrs Williamson. When he did eventually marry, at the age of forty-eight, the marriage was not a success.

Maybe, despite the certainty of Mr Delamotte, the lot had not been 'perfect' after all.

It is not God's fault that we can't hear him; it is ours for not having prayed hard enough to receive the 'perfect lot'. A 'bad lot' ensued when the prayer had not been sufficient to enable God to speak. As the statistician F.N. David notes, 'This was presumably the way out which is found necessary in all these procedures to allow for cases where the result of sortilege is not pleasing.'

*

In ancient Athens, political office was awarded by lottery (although there was a qualification to take part), just as jury selection is done today. Romans parcelled out land by lot to veteran soldiers, as Jesus's crucifixion clothes had been distributed, so that the most fertile would be allocated by chance.

Orville and Wilbur Wright tossed a coin to decide who would get the first opportunity to engage in powered flight. A toss of a coin determined the order of James Watson's and Francis Crick's names on their paper describing the structure of DNA.

In shipwrecks, when cannibalism resulted, deciding who was going to be eaten was legal if it was made by lots, rather than by position or rank or power. After the shipwreck of the *Mignonette* in 1884, the cabin boy Richard Parker was killed and eaten by his three surviving shipmates after several days out on open sea and all provisions gone. The survivors were rescued soon afterwards, returned to England and put on trial. The sentence of death was commuted to six months' imprisonment – which reflected the popular sympathy for the sailors – but the conviction was based on the sailors not having held a lottery to decide who was to die, even though Parker was weak and ill as a result of drinking seawater.

A fair lottery prevents corruption, ensuring that the distribution – of goods, of justice, of decision-making – is beyond human interference. No one is to blame for the selection or may take credit for it (it is disposed of by the Lord). If a lottery is the fairest system *in extremis*, if that is how we should choose which of us after the shipwreck should be eaten, or which of the hundred ministers should visit the plague house, why not use it in the normal run of things, such as to decide, as Wesley did, whether or not to marry Miss Hopkey?

As I write this, in 2021, there's a continuing pandemic of a coronavirus, Covid-19. Italy was the first European country to be hard hit early in 2020. In the first wave of sickness, news commentators often mentioned the 'difficult decisions' that Italian hospital administrators and doctors had to make. By 'difficult decisions', they meant that as a consequence of limited resources, patients were triaged and treated according to an estimation of their survival chances. 'It's a lottery,' the commentators often said. When, of course, it wasn't. A lottery would have been fairer, and more traditionally Italian.

The modern lottery is an Italian invention. In its first iteration, the lottery *blanco*, players paid a fee and wrote their name on a slip of paper along with a 'device' – usually a prayer, but it might be a proverb or a verse – which was registered with a corresponding unique number. The slips of paper were put into an urn or leather sack. A second urn or sack held an equal number of slips of paper, the majority of them blank, a few with a designated prize. At the drawing, a child would pick out a name slip, pass it to an official who would read out the device and the number; then the child would draw a slip from the second urn, which the official would announce as either a blank or a prize.

Allowing each player their own public reckoning with fortune could take a long time. Lottery tourists might swell the population of a town where a draw was taking place for a month, or longer. For example, a single white lottery in Hamburg in 1612 took fifty-seven days to be drawn.

*

The adventurer and gambler Giacomo Casanova came to Paris in January 1757, having escaped from a Venetian jail, a chancer on the make. He was resolved not to associate with cabalists and Rosicrucians and Freemasons, the 'dubious company' that had got him into trouble in Venice. A friend had given him an introduction to the Controller-General of Finances, Jean-Nicolas de Boullongne, along with the advice that Casanova should think up a money-making scheme that was not too complicated.

> I had to pay my court to those with whom the Blind Goddess made her abode. I saw that to accomplish anything I must bring all my physical and moral faculties into play, make the acquaintance of the great and the powerful, exercise strict self-control, and play the chameleon to all those whom I should see it was my interest to please.

When they met, Boullongne asked Casanova if he knew of a way to raise twenty million livres for the Royal Military School, without 'burdening the State or embarrassing the royal treasury'.

Casanova assured Boullongne that he did, and that he would shortly reveal what it was, on the condition that the Controller-General was able to guess what it might be. 'I left him full of gratitude,' Casanova writes in *The Story of My Life*, 'but at a loss to find a way to increase the King's revenues. Having not the faintest notion of finances, no matter how I racked my brains all the ideas which came to me were only for new taxes, and since they all struck me as odious or absurd I rejected them all.'

> I went to walk in the Tuileries, reflecting on the fantastic piece of luck which Fortune seemed to be offering me. I am told that twenty millions are needed, I boast that I can furnish a hundred millions without having any idea how to do it, and a famous man, thoroughly experienced in business, invites me to dinner to convince me that he already knows my plan. If he thinks he can worm it out of me I defy him to do it; when he has imparted his own plan to me I shall be at liberty to tell him whether he has guessed mine or not, and if the thing is within my comprehension I will perhaps say something new; if I don't understand a word of it, I will maintain a mysterious silence.

The following day, he was introduced to a group of politicians and financiers as a friend of the Minister of Foreign Affairs and the Controller-General. Casanova remained silent as his companions discussed the frozen Seine; the interrogation of Damiens, whose assassination attempt on Louis XV had coincided with Casanova's arrival in Paris, and the five million his trial would cost; and the good choice of Monsieur de Soubisse as the new commander of the army. 'This subject led to the cost of the war and to the resources for maintaining it and the country.' After an hour and a half of this, of Casanova's oracular silence amid the chatter of powerful men, he was finally led into a study, passed a folio notebook, and was told, 'There is your plan.'

It was a proposal that had been put forward by the Calzabigi brothers from Genoa for a national lottery, based on their home town's model for selecting public officials. The form of lottery devised in Genoa in 1610 shortened the previous lengthy procedure. This is the lottery that we're familiar with, that my neighbour Barbara plays when it reaches a rollover, the names and devices replaced simply by numbers.

Casanova agreed that this was indeed his plan. He was told that it was supported by the king's favourite mistress, Madame de Pompadour, but some members of the King's Council were worried, fearing losses. Casanova, the great improviser, said that would be no bad thing. He advised offering a guarantee of one hundred million livres. 'If the king loses a great sum at the first drawing, the success of the lottery is assured. It is a misfortune to be desired.'

Casanova became the front man of the Calzabigi brothers' version of the Genoan lottery, which was an immediate success. He had the luck to have the right introduction to the right person at the right time, and his luck is also now historical – the Calzabigis, whose idea this was, are now largely forgotten, and Casanova is credited with having brought the Genoese lottery to France, when he merely had useful contacts, the effrontery to claim great things, a moment of opportunity and a huckster's instinct to exploit it.

The nationwide lottery for the Loterie de l'École Militaire was drawn once a month by a blindfolded child – signifying purity, incorruptibility and, perhaps, God's grace – who drew the five winning numbers out of ninety from a revolving cylinder called *la roue de fortune*, 'the wheel of fortune'.

The lottery became a machine for giving hope to the poor or, rather, for selling hope to the poor. As one revolutionary critic put it, it was 'a scourge invented by despotism to silence the people in their misery, deluding them with hopes that only aggravated their sufferings'. The equivalent criticism from the reactionary right wing was that it threatened social order, 'these people drawn from the dust... are going to place themselves among the nobility. Only their money, and not their merit, elevates them to charges and functions which require talent and sensibility, which one ordinarily receives only by a gentle education.'

Despite Gataker's efforts to divide lots into sacred and profane, the categories refused entirely to separate. Like lines that will not straighten, they bend towards each other and merge, the random event and God's will.

There is however the perpetual problem of fairness, of protecting the pure channel for God's speech. Lotteries are open to manipulation, consciously or not. The mechanism for deciding the Vietnam draft in 1969 was inadvertently skewed towards the later months of the year: twenty-six days in December were draft days; if your birthday was in December and you were the right age without the right contacts to find means of deferment, the odds were you were drafted.

'Random' draws in football tournaments have often been gamed by certain balls being warmed immediately before the draw, so whoever was 'blindly' picking the numbered balls that corresponded to the teams could select pairs comprising one warm ball and one cold, to keep tournament favourites apart. And one wonders quite how involved Mr Delamotte was in the procedures of Wesley's appeal to the Searcher of Hearts in the matter of Miss Hopkey. It was Delamotte who drew the lot, 'Think of it no more', while presumably Wesley was in deep prayer to make it perfect. Delamotte was Wesley's companion and associate, who might have seen his own role as keeping his friend on the unmarried path, even, for example, nudging God's opinion in a certain direction ('only in the presence of Mr Delamotte').

*

Lotteries are also open to manipulation by the mathematically minded with deep pockets. The principles of how to game a lottery were established by the writer Voltaire (the nom de plume of François-Marie Arouet) and the mathematician Charles Marie de La Condamine. The collapse of John Law's 'Mississippi Bubble' in 1720 had devastated the French financial system. Government bonds, usually the safest form of investment, plummeted in value. To recompense bond-holders, and raise money for the state, Louis XV's then Controller-General of Finances, Michel Robert Le Peletier Desforts, brought in a lottery intended to drive up the prices of bad bonds and restore faith in state finances, and allow greater credit. Bond-holders were invited to buy a ticket for every bond they owned, at a thousandth of the bond's value. A winning ticket would gain its holder the face value of their original investment, and a jackpot of 500,000 livres (sufficient to make the winner rich for the rest of their life).

There was an immediate appetite for tickets, but the mistake Le Pelletier-Desforts had made was that any ticket, regardless of the size of bond it originated from, had an equal chance of winning. Each bond worth 1,000 livres allowed the purchase of a lottery ticket for 1 livre – as opposed to, say, a bond for 100,000 livres costing its holder 100 livres a ticket. Condamine had the idea to buy up a large proportion of the small bonds. He and Voltaire set up a syndicate to invest money in bonds and buy as many lottery tickets as they could, and as many as friendly notaries might allow. Aspirational players would continue to write the traditional 'devices' on their tickets; Voltaire wrote phrases such as 'Here's to the good idea of M.L.C.!' and 'Long live M. Pelletier-Desforts!'

Before the lottery was shut down two years later, Voltaire's and Condamine's individual shares in the syndicate profits were each about 500,000 livres. Voltaire added to his wealth by shrewd investments. Condamine made a subsequent career of measuring things (see Chapter 9).

*

Barbara's odds for winning her rollover jackpot are 5.3 million to one. She might find hope in the case of Joan Ginther, who has won four multimillion prizes on the Texas scratch-card lottery, for a total of $20.4 million. The odds of doing this are about 18 septillion to one. (For comparison of scale, there are approximately one septillion grains of sand on Earth.) 'She's obviously been born under a lucky star,' said a spokesperson for the Texas Lottery Commission, who might or might not know that there are approximately one septillion stars in the universe. More than one newspaper article about Ginther has declared her 'The Luckiest Woman on Earth'.

Ginther grew up in Bishop, Texas, the daughter of the town doctor. Three of Ginther's wins came from lottery tickets bought in the same convenience store in her home town, the other at a shop in nearby Kingsville. Between 1993 and 2010, she had a practice of coming back to town twice a year to stay for about a month in a motel near the Times Market, where she bought her tickets. The most in-depth examination of Ginther's success was written by the appropriately named Nathaniel Rich in *Harper's* magazine. Residents of Bishop told Rich how well liked Ginther is, how she gives out lottery tickets to anyone she sees, as well as cash to people in need. 'She mills around talking to people. People go there in the hope that she's handing out tickets. She says, "Hi, my name is Joan. Would you like a ticket? I'm a millionaire and I buy tickets and hand them out to people to see if they have any luck, too."'

Ginther majored in Mathematics at the University of Texas at Austin. After graduation, she studied at the University of Stanford's School of Education, where she completed a PhD in statistics. She then taught maths at a community college in San Jose, California, probably until the mid-1980s.

The Texas scratch card works in three-month-cycles. It doesn't serve anyone's interests if the big prizes are won at the beginning of the cycle. As with Barbara's rollover, holding back the big prizes increases the jackpot, creates natural publicity, raises consumer enthusiasm and increases everyone's profits.

This is how to win on the lottery: buy tickets late in the cycle; find some information about where winning tickets are likely to be; buy them at a relatively out-of-the-way place where you can maintain a monopoly; and buy them in bulk. Wenxu Tong, who runs a syndicate of lottery ticket gamblers or, rather, investors, in one month was said by a lottery agent to have won $280,000, on an investment of $200,000.

Five perfect shuffles of a physical deck of cards produces a random distribution. Nothing human-made can ever be truly random. A digital algorithm can at best generate a pattern that doesn't *appear* to be predictable.

The Times Market is now closed. Ginther seems to have found the publicity unwelcome and there are no reports of her returning to Bishop, or of further lottery wins. She remains, quietly, in Las Vegas. I imagine her walking along casino floors on the Strip, the Mirage, the Bellagio, Treasure Island, maybe with a hand-held computer in the pocket of her jacket (I imagine a red or purple hoodie, with a casino logo printed on the back), working out a way to beat the machines.

*

According to the Talmud, 'Four things annul the decree that seals a person's fate; namely, alms, prayer, change of name, and change of deeds.' The approved way to change one's name was to find the replacement by turning to a random page in the Bible.

There are several messages here: the importance of action, that it is right to give to charity and to be humble, and that change is in itself a good. Submitting to randomness can break the bad patterns. It's using luck, but calling it something else.

Kabbalistic literature often derives from this sort of idea. The *Sefer ha-Temunah* (the 'Book of the Image'), from thirteenth-century Spain, describes the central Jewish holy book, the Five Books of Moses, the Torah, as a body of spiritual letters which, although materially unchanging, presents different appearances in different cosmic aeons. Gershom Scholem writes that,

> In effect, therefore, each aeon, or *shemitah*, possesses a Torah of its own. In the current *shemitah*, which is ruled by the divine quality of *din*, stern judgment or rigour, the Torah is read in terms of prohibitions and commandments and even its most mystic allusions must be interpreted in this light. In the coming aeon, however, which will be that of *rahamim*, divine mercy, the Torah will be read differently, so that in all probability 'what is prohibited now will be permitted then'.

A book of changes – which also happens to be the English translation of the title of the *I Ching*, the ancient Chinese manual of divination.

Lao Tse and Confucius were both students of the Chinese *Book of Changes*. The Taoist view is that every event in the visible world is an 'image' of an idea in the unseen world. The events in heaven, the suprasensible world, take place before those on Earth.

The psychologist Carl Gustav Jung in his foreword to the Richard Wilhelm translation refers to 'the living meaning of the text', 'the living soul of the book', 'a spring of living water'.

> The axioms of causality are being shaken to their foundations: we know now that what we term natural laws are merely statistical truths... If we leave things to nature... every process is partially or totally interfered with by chance... The Chinese mind, as I see it at work in the *I Ching*, seems to be exclusively preoccupied with the chance aspect of events... There is something to be said for the immense importance of chance. An incalculable amount of human effort is directed to combating and restricting the nuisance or danger represented by chance.

I am not going to attempt to understand 'the Chinese mind'. I have promised myself to limit my investigations to the cultural traditions that have shaped me, or which I have chosen to be influenced by. I had also promised to visit my friend Carl, aka 'the Money', in Los Angeles. We are going to drive to Las Vegas together, where I will put some of my luck researches to the test. I called him up. The trip has to be delayed, I told the Money. I have an algorithm.

He scoffed at this. I told him that it was absolute, that the randomiser had placed the Vegas chapter at the end of the book. It's out of my hands, I said.

I do wonder if there is a kind of cowardice here. I am out of practice at poker, which is how I will be testing my luck, and Las Vegas is not a good place for a poker player to be when they are out of practice or out of form.

Instead of going to Las Vegas, I will be writing next about superstition, because that's what the algorithm has instructed me to do. And if I've learned anything in this chapter it is that luck is to be found where the chance element interrupts the stasis, the unknowing dart that kills Dido, the randomly derived change of name that allows the individual to move forward, Casanova coming into town with a letter of introduction. Also, it will give me time to play some more poker, and maybe to pursue further researches into the mysterious Joan Ginther.

I went to the supermarket to buy lottery tickets. Distracted, walking towards the entrance, I bumped into an abandoned trolley. In it was someone else's list, a previous shopper's. I picked that up and followed its instructions as if they were my own: I bought tofu, soy sauce, rice noodles, avocados, carrots, vegetable oil, flour and sparkling water. We are already well stocked with vegetable oil and flour, and I don't like sparkling water, but, as I'd slightly pompously informed the Money, a random procedure is its own moral imperative.

I also spent £5 on scratch cards and £5 on lottery tickets. This is much less than Joan Ginther's outlay, or Voltaire's. But more than Barbara's.

Barbara didn't win on the lottery. There was a baffled expression on her face when she told me about it. Neither did I. The first scratch card, which cost £1, gave me a 'bonus' scratch that won me £1. The rest were all blanks. For an outlay of £10, all I won was that first £1.

lottery expenditure −£10
income £1

CHAPTER 6

My Lucky Underpants

It is especially in games of chance that the weakness of
the human mind and its tendency toward superstition
manifest themselves. And it is much the same for
people's behaviour in all those areas of life where
chance plays a role. The same prejudices govern them,
and imagination dictates their conduct, blindly giving
birth to fears and hopes.

Pierre Rémond de Montmort, *Essay on the
Analysis of Games of Chance*

'tis commonly allowed by philosophers that what
the vulgar call chance is nothing but a secret and
conceal'd cause.

David Hume, *Treatise on Human Nature*

net total –£309

My father took me to my first casino. In a failed attempt at
father–son bonding, we were taking a boat down the Eastern
Atlantic seaboard from Montauk, New York to Norfolk,
Virginia. I was about twenty-three, which meant he would
have been sixty-one. We docked overnight at Atlantic City.
He led me on to a casino floor and gave me a hundred-dollar

bill. 'Come back to the boat when you've lost this,' he said. He was teaching me a lesson about gambling, a practical demonstration of his 'fool and a t'ief' theory.

It became a point of Oedipal pride that I would not lose the money. I made my way to the blackjack tables, because I knew the rules of that game better than any other, and also because I'd picked up that it was the casino game which gave the house its smallest edge. I used to play 'pontoon' at school, aged twelve or thirteen. On a trip to Hampton Court, a few of us had got deliberately lost in a side avenue of the maze, where we played for pennies.

At the casino in Atlantic City, an old man who looked just like Hemingway, if Hemingway had allowed himself to become old and shrunken, was playing two boxes of blackjack at the same time. He was the only one sitting at the table and the dealer would deal him two hands and he would make two sets of bets, and I thought, Wow, this guy is good, I'm going to watch him and learn from him, he will be my true, preferable, father, from whose wisdom I shall profit. (I was often looking for preferable fathers in those days.) I had assumed that he was playing two hands at a time to enable him to win twice as quickly. It soon became apparent that the opposite was true. I watched for maybe ten minutes and in that time old Hemingway stoically burned through several hundred dollars.

I went to another table, where I played cautiously and maybe it was beginner's luck that enabled me to nearly double my stake. I returned to the boat and handed my father the $180. I'd hoped to return his stake and be allowed to keep the $80 profit. He told me to keep all of it. There was something like pride on both sides, and maybe also some disappointment from his.

I was wearing a short-sleeved shirt of alternating dark and light green stripes, which I had previously thought of as my carnival-in-Rio shirt, and which I now immediately designated as lucky.

*

Nearly a quarter of a century later: I lived in London, my father was in Manhattan. As his health declined, I would visit him more often. I would ferry him to doctors' appointments, try to intervene in the wars he was fighting with his wife, my stepmother. A couple of times, in 2006 and 2007, I went to see him on my way to Las Vegas.

'What you doing there?' he'd ask. 'Gambling?'

'No,' I'd tell him, patient and a little superior, 'playing poker.'

And then the next time, we'd have the same conversation. I'd try to explain to him that in casino table games you play against the house with its inbuilt advantage; in poker, you're playing against other people.

'What you doing there? Gambling?' he'd say again. It just wouldn't take with him that poker is a separate species from gambling. And of course it isn't, not quite.

Each time I visited him, I learned a little bit more about some of the stories of his past. And each time, he'd tire of the telling. What was the point? I wouldn't really understand. His wife sometimes urged him to attend get-togethers of Holocaust survivors. He sneered. 'And what for? So we can congratulate each other?'

But I'm getting this wrong. In 2001, when he was seventy-nine, Joe Flusfeder suffered a stroke. With an enormous effort of will, he was partly successful in regathering himself, and while new language pathways established themselves, for the remaining years of his life he struggled with words, both comprehending and producing them. Many of his thoughts would be locked in behind a wall of unavailable language.

He would order calamari when he wanted a cappuccino. He had trouble with his pronouns, with the effect that he sounded quite cantankerously camp, referring to men as 'she'. Multiway conversations lost him. His most commonly used phrases were 'Absolutely!' when he was, as if genially, agreeing to a suggestion he hadn't understood, and an exasperated 'Forget about it' when he had failed, again, to say what he had intended to say.

So we wouldn't have had this conversation about attending a group for Holocaust survivors in 2006 or 2007. That would have come earlier, before 2001. In 2007, he was no longer talking to his wife, but he would have asked me if I was gambling.

*

July 2007, Day 1C of the Main Event of the World Series of Poker. I had played the tournament for the first time the year before, and got about halfway through Day 2 when my extravagantly unnecessary bluff failed and I was knocked out by a thin affectless New Yorker, who was wearing a red T-shirt with the inexplicable words 'Mr Wonderful' printed in white on the chest. It had taken me the best part of the year to get over my ensuing fury and shame. Two, Dostoevskian, mistakes had contributed to my tournament suicide. The first had been my increasing irritation at a conversation going on between two of my neighbours, a solid poker pro called Gary and his relentless interlocutor. I'd learned early that the pro's name was Gary because his interlocutor used it at the end of each of his very dull sentences, and the sound of the name had been echoing in my head in a torturing sort of way. The second mistake was that Mr Wonderful had only recently been moved to our table, so would not have been familiar with the image of myself that I had constructed, of a conservative, nitty player who would probably only bet for value.

And now, determined not to lose my composure, I was back in the biggest poker tournament in the world, with its $10,000 entry fee. I was one of over 6,000 entrants, of whom about 10 per cent would receive a prize, up to $8.25 million for first place.

After the initial nerves had subsided, the playing styles and, therefore, characters, of the players around the table had become evident. The Canadian to my right liked to make strong bluffs, an aggression which you had to answer with equivalent aggression. Jeff from New Jersey, who was sitting two to my left, had a powerfully inflated opinion of himself, which had aroused the ire of Israeli poker pro David Levi down the other end, who had spent much of the last three hours making fun of him. Across from me was Cristina, from Alaska, a cheery lady of advancing years, who was showing more of her cleavage than is customary in a heavily air-conditioned room. And on her left was a young Norwegian, who played what was then a fresh style, ultra-aggressive, raising everything, almost regardless of his cards.

The stealing Scandinavian raises, I call from the big blind. The flop comes down king of clubs, ten of diamonds, two of spades. I've got the ace and king of spades in my hand. I hadn't reraised preflop in order to disguise my strength. I could check here, raise the Norwegian's probable continuation bet. Instead I lead out betting. I can't entirely remember my thought processes, but I think I was intending it to look like a weak, donkish probe bet, as if I was testing the strength of the Scandi's hand to see if my pair of tens was good.

Cristina folds. The Scandi calls, which tells me nothing.

The turn is the ten of spades. I've now got top two pairs, with the top kicker and the nut flush draw. I bet, he raises, putting me all in. What has he got? He is capable of raising preflop with a very wide range of hands. I want him to have king-queen. I don't think he does. His hand must be stronger than that. My agony is transparent. I ask for time. I need it.

Around us are the sounds of the tournament, the relentless clacking of chips, the murmurs of conversation, a dealer shouting *All in on table one!* for the TV vulture cameras to gather for another tournament death. From a table further away, near the middle of the room, a player who's probably in a similar situation to me calls out so plaintively, *Don't send me home. I don't want to go home.*

I don't want to go home either. It's not exactly that I like being here, in many ways it's intolerable: the constant making of difficult decisions; the whining of Jeff from New Jersey; Levi's relentless banter about Jeff's sexuality, about Cristina's breasts; the constant necessity of assertion and reassertion – but I don't want to leave it either. The phrases 'tournament life' and 'tournament death' are not empty ones. I do feel alive. I want to stay that way.

Has he got a pair of twos in his hand? King-ten? I'm in trouble. Why ever did I call with AK? I should have raised, long ago, when I might still have had some sort of control. If I've convinced him I've got a ten in my hand, then surely he's got to be super-strong? Pocket kings? Unlikely, given that I've got a king myself. I'll probably need to get lucky to win this hand, but someone has to stand up to him. Why does someone have to stand up to him? And even if they do, why must it be me? There's no reason why I should be the table policeman. I tell myself to accept the loss of the chips that have already gone into the pot. I'll have a below-average stack but still be alive. All I can really beat is a bluff. Prudence tells me to fold. I call.

Scandi shows his pair of twos. He's got a full house. My flush draw is worthless. With one card to come, only a ten or a king can save me. I stand up. It feels all over, inevitable. There are four cards left in the deck that can save me, two kings, and two tens, to give me a superior full house. *I don't want to go home*. My odds of winning this hand are less than 10 per cent, or in probability terms, below 0.1. There should be some forces to call on but I don't know what they are.

This isn't my father in Warsaw or Siberia or Monte Cassino. It isn't even Ashley Revell at the Plaza Casino. In most ways this doesn't really matter. I have a hotel room downtown, some money in my pocket, a plane ticket home, a family that will be pleased to have me return, a professional life that brings me some income.

The dealer lightly taps the table, the beat in time before the river card is turned over.

This moment has already stretched itself out into an unfeasibly long duration, but it is also precarious: it, and I, might be about to snap at any single instant. I marvel that I am still standing, that Vicky in the seat to my left is still about to take a sip of tea.

Cristina from Alaska says to Levi, 'Are you going to put your head on my chest or what?'

I've come a long way for this, and prepared well. I'd been winning small tournaments in London cardrooms, my form, as they say (and 'form' in this context means luck, or, as poker players call it, 'variance'), had been good. I'd got to town in plenty of time to allow my body some relief from jet lag. French Eric and I, before he had to return to London, had talked through strategy. He'd railed me in a couple of tournaments, giving feedback on my play, my manner at the table, the decisions I had made, the way I had implemented my decisions.

My return flight is in four days' time – I'd allowed for the possibility that I'd make Day 2 again – but I'll have to bring it forward, because I don't want to be here anymore.

The dealer turns over the river card. It's a ten.

I've hit one of the four cards left in the deck to give me the better full house. I've won the hand and doubled up.

When the night is over, I'm still in the tournament. It's three in the morning and I share a taxi back to my hotel with Jens from Denmark, who had been knocked out in the very last hand of the night. 'I had king-queen suited in the cut-off,' he says, 'and only fifteen thousand in chips. I figure at least I can get the blinds with this hand. So I go all in and get called by the big blind with aces.'

He tells it plainly, giddy with tiredness and half-crazed with disappointed adrenaline. We've both played poker for thirteen hours and the only difference between us is that I have chips and he doesn't. Or to put it another way, I got lucky and he didn't. 'So how do you feel?' I ask him. He stares at me. 'I don't know.'

I wouldn't let Jens pay his half of the cab fare. I waved away the fifty-dollar bill he tried to offer me for his share of a twenty-five-dollar ride. I think it was because I was feeling guilty for my greater luck, but also because, after the hours of concentrating and reckoning, my brain couldn't quite manage the calculation. The driver had watched the exchange in his rear-view mirror. Jens lurched off out of the cab and I paid the fare.

'He lose, yeah?'

'Yeah.'

'No wonder.'

It took me a moment to work out what he was saying. The driver was disapproving of the fifty-dollar bill that Jens had been offering and which he had put back into his pocket.

*

You never see fifty-dollar bills in a casino. They're unlucky. Everyone in Vegas knows that.

One theory is that it's because of stock car racer 'Little' Joe Weatherly. Weatherly had a fear, close to phobia, of getting trapped inside a burning car. He wouldn't wear a shoulder harness or use a window net restraint, so when his car sideswiped a retaining wall at the Riverside Raceway in Southern California in 1964, his head smashed against the wall through the opened window, killing him instantly – and two fifty-dollar bills were supposedly found in a shirt pocket.

A similar theory has the violent death being that of the gangster Ben 'Bugsy' Siegel. Pioneer mobster of Las Vegas, associate of Meyer Lansky and Charles 'Lucky' Luciano (whose nickname was derived from a childhood mispronunciation of his surname), Bugsy was gunned down at his home in Beverley Hills in 1947. He, supposedly, had three fifty-dollar bills in his shirt pocket.

The most popular explanation is that the fifty-dollar hex came about because the banknote carries the image of Ulysses S. Grant. Grant was the first four-star general in American history, the man whose military genius inspired the Union side to victory in the Civil War, a two-term president from 1869 to 1877. He was also an alcoholic who, before returning to the army after the Civil War had started, had failed at farming and real estate. Grant was widely seen as one of the worst presidents in American history, whose tenure, in the so-called 'Gilded Age', allowed a flourishing of corruption and profiteering, and a huge rise in poverty and disease and social disorder. After bailing out his son's failed stockbroking company, he died broke in 1885. His real name was Hiram Ulysses Grant. Ulysses had been chosen by the drawing of lots; the 'S' was a mistake made by his nominator on his West Point application form.

Ferdinand Ward, the swindler who ruined Grant financially, also played poker with him. 'I think the game appealed to him because he had to bring to it many of the same qualities which caused him to be determined to "fight it out along this line if it takes all summer",' Ward wrote, referring to a quote of the general's. 'The possibilities for ambuscades, masking of batteries and sudden sorties in the great American indoor game appealed to him immensely.'

The journalist George A. Townsend related Grant's poker style to his ability as a military commander – even if using that approach in a cardroom would more or less guarantee that he was going to go bust. 'You know how a man of Grant's temperament would bet. The first wager he made would be with all he had for all on the cloth. "All the downs" was his favourite bet. He did the same in war... He felt the spirit of the game and played for big victories and promotions.'

Here's one of his maxims,

'The art of war is simple enough. Find out where your enemy is. Get at him as soon as you can. Strike him as hard as you can, and keep moving on.'

And here's another,

'One of my superstitions had always been where I started to go anywhere, or do anything, not to turn back, or stop until the thing intended was accomplished.'

Reluctantly, I confess to sharing this second one. In my early twenties, I was on an overnight ferry from Crete to Athens, and had fallen in with a Breton hippie named Jacques. We had spent the night drinking ouzo and bonding and annoying others. The ferry docked at about seven in the morning and we were hungry. Wandering the harbour streets of Piraeus, we finally found a café that was open. About to go in, I saw another place on the next corner and thought we ought to investigate it, to compare. The second place was closed and I was about to return to the open one but Jacques stopped me, saying, 'No. We must never turn back.' This felt over-literal to me, and I resisted it, but he was adamant and we had become friends so I stuck to his code and we walked for another hour before we found anywhere open for food. I have stayed true to this superstition ever since, maybe not with the same rigour as Jacques, but I have kept it as one keeps the gifts of friendship.

*

I grew up trying to resist superstition. My mother would never walk under ladders or open an umbrella indoors or put on her right shoe before her left. If she spilled salt, she would immediately throw pinches of it over her left shoulder. Usually mild and softly spoken, she once shouted at my sister for walking over me when I was lying on the floor because, she said, that could stop me from growing. You could not place keys on her table. A shirt could not be worn while a replacement button was being sewn on to it.

She had learned these rules from her own mother, with whom we lived after my mother had left her American marriage behind to return home to London with her six-year-old son and eleven-year-old daughter. My grandmother had been born deep in the nineteenth century in a Polish shtetl. She mistrusted such manifestations of modernity as telephones, refrigerators and escalators. Even in her eighties, she would walk up a Tube station emergency staircase rather than use 'the moving stairs'. Her life was governed by magical rules, many of which she had passed down to her daughter. Once, when we were feeding the ducks in Springfield Park in Clapton, a benevolent passer-by asked my age. My grandmother told him I was four and grabbed roughly at me to make me quiet, before, outraged, I was able to correct her. When we got back to the flat, I showed my mother the bruise on my arm. She was less sympathetic than I would have liked. She told me that my grandmother believed she was protecting me from 'the evil eye', the *ayin hara*.

I questioned the concept of the evil eye. What makes eyes so fearsome? All they can do is look.

In this way, I was allying myself with my faraway father, who saw my mother's family, the Tessers, as ignorant peasant types, prey to superstition, dwellers in a dark, slightly contemptible past. The Flusfeders were from Warsaw; he had been brought up to be a cosmopolitan city boy, a modern. This is what I would be too.

And yet, these things leave their marks. My grandmother lived in a world of invisible threats, of demons whose interposition into our lives she fought every moment of the day to resist, to close up any channels along which they might enter. Everywhere around us was the evil eye, multiplied, countless. Even though I had chosen not to believe in this, the threat, the warning, the image of it, was persuasive and seeped in. A self-conscious child, I was already anxious about other people's gaze, which I experienced as touch, even assault.

The notion of the evil eye is necessarily social. It presupposes a crowded world, of different ranks of achievement and success, and luck, populated by those who have and those who are envious. 'When any one looks at what is excellent with an envious eye he fills the surrounding atmosphere with a pernicious quality and transmits his own envenomed exhalations into whatever is nearest to him.' This is from the Byzantine writer Heliodorus. It could come from Polish folklore, the tale of the father with a loving heart, but afflicted with the evil eye, who blinds himself to protect his children;

or from ancient Rome, where a common phrase for an illness without obvious cause was that the victim 'had been looked on'.

The same idea is there in the Old Testament,

> Now when David returned, after he slew the Philistine, the women came out of all the cities of Israel, singing and dancing, to meet king Saul, with timbrels of joy, and cornets.
>
> And the women sung as they played, and they said: Saul slew his thousands, and David his ten thousands.
>
> And Saul was exceeding angry, and this word was displeasing in his eyes, and he said: They have given David ten thousands, and to me they have given but a thousand, what can he have more but the kingdom?
>
> And Saul did not look on David with a good eye from that day and forward. (1 Samuel 18:6–9)

The Talmud instructs us not to gaze at someone's field of standing grain, lest we damage it with an evil eye. Nor do we count people, which is the tradition my grandmother was working in, when she refused to state my actual age. Rabbi Elazar said, 'Whoever counts a group of Jews violates a negative *mitzva* ["duty" or "commandment"], as it is stated: "And the number of the children of Israel will be like the sand of the sea, which cannot be measured."' (Hosea 2:1).

The evil spirits must be kept uninformed. We might remember here that Izio Flusfeder avoided two occasions when he might have been counted by the Nazis: that moment in the round-up in Warsaw where he was the lucky beneficiary of another's wisdom and walked straight through the square and out again; and when the German authorities refused to put him on their list of Polish refugees in Kovel.

Back in the shtetl, the world my grandmother was born into, little visible preparation for childbirth would be made. After childbirth, before the baby was named, she would have been in special peril from Lilith, Adam's first wife, who wants to snatch all babies to make up for her own demon children who are killed daily. If the baby laughs during the night, she must quickly be slapped, because Lilith may be playing with her. Anyone other than parents or grandmother who looks at the baby must be countered with three spits. As the child gets older, the danger diminishes, but vigilance must be maintained. There can be no boasting of the child's accomplishments, or any act that might arouse envy; these are invitations to the evil eye.

When discussing a plan, or anyone's accomplishments, my grandmother would always say *kinehora*, 'let it be without the evil eye'. And if still anxious, she would spit three times.

This practice goes much further back than the nineteenth century. Pliny the Elder, in his *Natural History*, reported that 'we spit on epileptics in a fit, that is, we throw back infection. In a similar way, we ward off witchcraft and meeting a person lame in the right leg'.

Spitting is also, he notes, efficacious against leprous sores and eye infections, cancer and neck pains... 'let us also believe that any insect that has entered the ear, if spat upon, comes out. It acts as a charm for a man to spit on the urine he has voided, similarly to spit into the right shoe before putting it on, also when passing a place where one has run into some danger... If we hold these beliefs we should also believe that the right course, on the arrival of a stranger, or if a sleeping baby is looked at, is for the nurse to spit three times at her charge.'

The Greek word for superstition is δεισιδαιμονία (*deisidai-monía*), literally, 'fear of demons', so my grandmother was in her own way something of a classicist. The Latin *superstitio*, which means 'standing over', denotes an attitude of irrational religious awe or credulity – as in, 'Look at what *they* believe; but *we* are civilised, but *we* know better.'

'Superstition' is the 'other' to 'science'. But it's also the other to religion. And it's the rural other to the city, or the primitive other to modernity. As Pliny says, *If we hold these beliefs...* We have, he implies, a choice. A few generations earlier, in the late Republic, Columella's *On Agriculture* dismissed 'such religious observances of our forefathers' as the requirement to sacrifice a puppy before cutting hay or shearing sheep. Columella was presenting the modern science of agriculture – like my father talking about his mother-in-law, or Columella's contemporary Cicero ridiculing the still popular belief that it was a bad omen when two yoked oxen simultaneously defecated, the religious practices of the past become the superstitions of now, to be despised by the rational.

The sociologist Max Weber said, 'there are no mysterious incalculable forces that come into play, but rather that one can, in principle, master all things by calculation. This means that the world is disenchanted. One need no longer have recourse to magical means in order to master or implore the spirits, as did the savage, for whom such mysterious powers existed.'

Except, the world is not disenchanted. Not 'all things' can be, even 'in principle', mastered by calculation. The world is so dangerous, we are besieged on all sides, our chances so precarious, we need all the protection we can muster.

Montaigne wrote,

When one scale in the balance is altogether empty I will let the other tilt under an old woman's dreams: so it seems pardonable if I prefer the odd number; Thursday rather than Friday; if I prefer to be twelfth or fourteenth at table rather than thirteenth; if I prefer on my travels to see a hare skirting my path rather than crossing it, and offer my left foot to be booted before the right. All these daydreams are in credit around us and deserve at least to be heard. For me they are only carrying inanity, but they are carrying something. The weight of vulgar and casual opinions is more than nothing in nature; and he who will not suffer himself to proceed so far tumbles into the vice of obstinacy, to avoid that of superstition.

The superstition against Friday the thirteenth did not exist yet, but Montaigne seems to come close to it here. Fears of the day Friday and the number 13 had long preceded him; they weren't combined into a single object of dread until the early twentieth or late nineteenth centuries. Jesus had died on a Friday, because crucifixions in the Roman Empire, like executions in the United States in the nineteenth century, always took place on that day.

Thirteen is an incomplete number. There are twelve months in the year, twelve signs of the zodiac, twelve gods of Olympus, twelve sons of Odin, twelve labours of Hercules, twelve tribes of Israel. Adding one more unbalances things, makes them dangerous. Norse legend tells of the dinner party in Valhalla where Loki arrived as the uninvited thirteenth guest and took revenge on his snub by arranging a murder. And of course, the thirteen guests at Jesus's Last Supper also has something to do with the legend of unlucky thirteen.

Like Montaigne, the composer Gioachino Rossini regarded 13 as an unlucky number and Friday an unlucky day. He died on Friday, 13 November 1868.

Arnold Schoenberg shared these fears. He wrote, 'Indeed, I am not so well at the moment. I am in my 65th year and you know that 5 times 13 is 65 and 13 is my bad number.' He survived that year, but he too died on a Friday the thirteenth, in 1951, at the age of 76. (7+6=13, which I'm sure he had noticed.)

At least a composer, like a writer, is generally in charge of the immediate conditions of their working life. Schoenberg could banish the number 13 from the pages of his scores, he could introduce a 'complete' twelve-tone system that refused to allow in a thirteenth. He could abbreviate the name of the character of Aaron in his opera so that the title *Moses and Aron* contained only twelve letters and not thirteen.

It's sports people, for whom most aspects of their working life are out of their control, who exhibit the most extreme displays of superstitious behaviour. Magical thinking and displacement activities and customs and talismans and amulets and childlike transitional objects all mesh together into the pre-match 'routine'.

University of Kansas football coach Les Miles eats grass from the field before a game (it 'humbles me as a man... lets me know that I'm part of the field and part of the game'). If anyone touched former baseball outfielder Kevin Rhomberg he would have to touch them back straight away. Michael Jordan used to wear his lucky North Carolina college shorts under his regular Chicago Bulls shorts. Barry Fry, when he was manager of Birmingham City football club, urinated in all four corners of the dressing room to break a 'gypsy curse'.

The boxing trainer Freddie Roach always puts his left shoe on before his right, as Montaigne did, as my mother and grandmother did, as I do (although I call it 'habit'). Members of Manny Pacquaio's retinue were always seated in the same positions at ringside. Many boxers will not wash in the days leading up to a fight. Avoidance of sex is well documented:

Muhammad Ali, for example, would be celibate for at least six weeks before a fight.

Tennis is the most rife with superstition. In tennis, as in boxing, the player has no one else out there to protect them. Between the moments of action, as well as during them, there is so much time to think, and doubt, and lose the learned rhythms of technique, and to be afraid. Serena Williams bounces the ball five times before her first serve and twice before her second; she wears the same pair of socks throughout a tournament. Andre Agassi forgot to pack underpants for his first-round game in the 1999 French Open. He'd lost his first match there the previous year but won this time; so he didn't wear underpants throughout the tournament all the way to the title. Goran Ivanisevic watched *Teletubbies* each morning on his way to winning the 2001 Wimbledon championship.

Rafael Nadal has an unvarying routine both on court and before he goes on, which begins with a cold shower, followed by the same ceremony of preparation: the bandaging of his knees, the taping of his rackets, the putting on of his bandana, bandaging of his fingers, followed by a series of explosive runs and jumps and exercises, while music plays. The routine is punctuated by frequent urinations. 'I find myself taking a lot of pees – nervous pees – just before a game, sometimes five or six in that final hour.' Before going out on court, he checks that his socks are exactly the same height on his calves.

Once on court, I sat down, took off my white tracksuit top, and took a sip from a bottle of water. Then from a second bottle. I repeat the sequence, every time, before a match begins, and at every break between games, until a match is over. A sip from one bottle, and then from another.

And then I put the two bottles down at my feet, in front of my chair to my left, one neatly behind the other, diagonally aimed at the court. Some call it superstition, but it's not. If it were superstition, why would I keep doing the same thing over and over whether I win or lose? It's a way of placing myself in a match, ordering my surroundings to match the order I seek in my head.

Saliva and urine, semen and blood – sports people, like alchemists and the acutely sick, are particularly concerned with bodily fluids. I've done my best to eliminate superstition. I'll walk without qualms under a ladder, but if I see keys on a table, I'll remove them to a less ominous surface as if, despite my modernity, my twenty-first-century self tapping away on a laptop with a smartphone in my pocket, there is still that nineteenth-century peasant inside of me, my grandmother spitting three times when she felt the evil eye upon her.

I'm writing this in my 'work shirt', which, like Nadal with his headband and water bottles, I associate with working. I don't associate the wearing of the shirt with the writing of especially effective or beautiful sentences, but I do associate it with moments of concentration on the activity of writing, separate from the other things that I do. It's a black T-shirt that has the words, in white, *ARCHER RECORDS PRESSING PLANT DETROIT since 1965*. There are many associations I have with this shirt, but chief among them, for the purpose of sitting at my desk anyway, is the Detroit ethos of work and process. If Detroit can produce cars and records and techno and Motown, then surely I can extrude my own words, sentences, chapters, books. It's not sympathetic magic, there's nothing superstitious about it. This is routine. Habit, like putting on my left shoe before my right.

The work shirt will eventually, like its predecessors, have to be replaced. It is already faded, but that's not the issue: I've worn work shirts well past the point where the material is ripped, and my elbows poke through. It's as if there's something intrinsic about them, an energy, like the power of a favourite song, or the charm that a talisman bestows, that eventually gets discharged.

I am careful to remove the shirt as soon as I stop working. This is partly a way to section off different parts of the day. I am done with work, I might return now into more social and domestic worlds. But there's also a protectiveness about this. Keep the charm fresh, protect it from the contaminations of the everyday. Make it last. Because maybe there's a finite amount of the stuff – as a friend of mine, a fellow cyclist, was told by the doctor attending him after he had been pulled relatively undamaged from under the lorry that had crushed his bicycle, 'Everyone has a little bit of luck in their lives and you've just had yours.'

You don't want to waste your lucky underpants by wearing them when you don't really need them. You don't want to travel charmed, every red light turning green, every doorman becoming your usher, wolves lick your hand, each obstacle bows beneath your feet – and then, you get there, to the job interview, the poker tournament, the rendezvous, only to find that your underpants have, as it were, become discharged.

*

In 2007, I got through Days 2 and 3 of the Main Event as well, and some way into the prize money.

Each morning, I followed the same procedure as the previous day. I had the same meal at breakfast, orange juice and an omelette with wheat toast, followed by coffee, and then I went to the outlet to buy another pair of green underpants, as I had done before Day 1. There was a pleasure in the pulling

136

open of the plastic sleeve, unfolding the underpants (green, always green, with some repeated spherical pattern going on), throwing away the new rustling sheet of tissue paper, and putting aside the pair of the previous day.

Everyone knows that green is an unlucky colour but I didn't care, it was my secret. I wouldn't have worn green colours openly, just as I wouldn't have eaten nuts at the table or carried fifty-dollar bills.

I wanted everything to be the same, the same procedures, the same food eaten, the same mood, with the same results.

In his *Tractatus Logico-Philosophicus*, Wittgenstein writes,

Proposition 5.1361
The events of the future *cannot* be inferred from those of the present.

Superstition is the belief in the causal nexus.

This applies to everything: extrapolating financial data to predict the future shape of the economy; supposing what life will be like after Covid-19 lockdown because of how things used to be before; bouncing tennis balls the same number of times during a match, to clear the mind, and allow the mechanics of the body to do their thing, unclouded by anxiety or self-consciousness – and because you won before when you bounced them those same five times, so surely that will help you win now; or going to the poker table having followed the same breakfast and underpants procedure, in the hope that the same luck will apply. (Have I mentioned that Day 4 fell on Friday the thirteenth?)

My attitude was more aggressive than on previous days. I was going to open up, look for opportunities. I was going to allow more room for luck in the effort to build my stack and make it into the properly big money. Of the 6,358 entrants, there were 337 players left. We were all guaranteed to win at least $39,000.

Nearly an hour had gone by when the small stack at the table went all in on three consecutive hands. The first two times, he took the blinds and antes unopposed. The third time he was called in middle position by a player with one of the larger stacks at the table. The action came around to me and I looked down at ace-king. Folding was out of the question. Calling was a reasonable option. So was going all in myself. Middle position might fold, and I would be heads-up against the short stack, who I was sure had an inferior hand to mine. Or middle position could call and I would have the chance – admittedly at the potential cost of my tournament life – to move from a below-average 275,000 in tournament chips to a flourishing 600,000 plus.

I went all in. The larger stack dwelled up and finally called. Time slowed again as we tabled our cards: my ace-king, the short stack's 10–9, the big stack's pocket jacks, which was a slight favourite to win the hand. The flop came down jack–9–2, which gave the large stack a set of jacks, and changed my chances of winning from 40 per cent to 1 per cent. I stood up. I gathered my jacket. The dealer waited, because that is what they do when deep into the tournament, building up the suspense, giving the bloggers and chip counters and TV cameras more time to do their things. The moment stuttered forward again. The turn card was a ten. The short stack now had two pairs, but that didn't matter, he couldn't knock me out, I didn't care about him or his hand. I now had a gutshot, I was down to a four-outer, just like I'd been on Day 1: a queen on the river would give me a straight, more than double up my stack and put me into contention. My odds had gone up to 9.52 per cent.

Photography by David Lloyd

I don't remember what the river card was because I didn't see it, except to register that it wasn't a queen, and I was knocked out, in 321st position, for a prize of $39,445. (Have I mentioned it was Friday the thirteenth?)

The prize money from the 2007 Main Event can't be counted in my running Luck tally. It's an outlier in my poker career, and it's too long ago. The sums must be of money won and lost in the light of the Luck 'journey' I'm taking while writing this book. All I can include is what I've made or lost while writing it. This is my own way of keeping score, maybe even of measuring the operations of my own luck.

While writing this chapter I've been playing, and winning at, low-stakes Omaha online.

online poker £109

CHAPTER 7

I Saw Dangeau Play!

The Prince de Ligne… frequented the Paris clubs and was celebrated for the *sang froid* he displayed in the face of huge losses. Day in, day out, he behaved in the same way. His right hand, which constantly wagered vast stakes on the tables, hung slackly. His left hand, however, was immobile, held horizontally across his right breast beneath his jacket. Later it became known, through his valet, that there were three scars on his chest – the precise imprint of the nails of the three fingers that had lain there so motionlessly.

Walter Benjamin, 'Short Shadows (II)'

Get it quietly.

Traditional poker saying

net total −£190

Writing to her half-sister the Raugravine Louise in 1695, Elisabeth Charlotte, the Duchess d'Orléans, asked, 'Is dancing out of fashion everywhere? Here in France, whenever people get together, they want only to play lansquenet. It is the game in vogue here. Young people no longer have any desire to dance.'

Maybe Madame, as she was known – she was married to the King's brother, the Duc d'Orléans, known simply as 'Monsieur' – was appeasing her half-sister's German Palatine Protestantism, which would be disapproving of decadent French Catholic ways. She went on as if anxious not to be misunderstood, or implicated in the fashion,

Happily for me I no longer like cards, for I am not rich enough to risk my whole fortune as other people do, and I have no taste for little stakes. They play here for frightful sums, and the players are like madmen; one howls, another strikes the table so hard that the room resounds, a third blasphemes in such a way that one's hair stands on end, and they all seem beside themselves and are terrifying to see.

It wasn't just 'the young people'. Madame's brother-in-law, Louis XIV, held three-hour-long *appartements du Roi*, three times a week, for the purpose of gambling. The queen, Maria Theresa, hosted a game in her quarters each evening, from eight to ten, when the King would arrive to take her to supper.

The Venetian Embassy in Paris ran four separate gaming rooms. By special permission of the King (capitalised: King Louis XIV, *le Roi*, always), the Duc de Tresmes, as governor of Paris, was allowed to use the Hôtel de Gesvres as a gambling house. Throughout this period, the state levied a tax on playing cards as well as on the *maisons de jeux* and the newly instituted *loterie royale*.

Gambling's status was *toleré mais non permis* – officially tolerated rather than permitted. But there were also hundreds of *tripots*, clandestine gambling houses, throughout the city, marked for the connoisseur by the lamps lit in front of them, which functioned as an unofficial municipal lighting system.

'Flaming pots set the Paris night ablaze,' noted the chief of police. And it wasn't just Paris: Bordeaux for example had more than two hundred *tripots* – and those were just the ones known to the police.

The Archbishop of Reims was said to have lost 40,000 livres (roughly equivalent to about £35,000 today) in less than thirty minutes while gambling in a carriage following the King's boar hunt. Mme de Sévigné writes of the queen having lost 60,000 livres before noon one day in 1675. A few years later there were reports of the King's displeasure at the news that his favourite mistress Madame de Montespan had lost four million livres in one evening of cards. He was relieved to learn that she had retrieved the situation by forcing her companions to play on through the night until she had recouped most of her losses.

Between 1643 and 1777 gambling was prohibited by thirty-two royal and parliamentary edicts, which gives an idea of how popular and widespread a pursuit it must have been in France, and how ruinous.

Events in the courts of Louis XIV (who reigned from 1643 to 1715) and Louis XV (1715–1774) are well documented. Gambling is everywhere in the chronicles of the time; in the memoirs of Louis de Rouvroy, Duc de Saint-Simon, and in the letters of Marie de Rabutin-Chantal, Marquis de Sévigné, you'll find more mentions of card games than food.

In a letter to her daughter, the comtesse de Grignan, Madame de Sévigné wrote,

They play extravagantly high at Versailles: the hoca is forbidden at Paris under pain of death, and yet it is played at court: five or six thousand pistoles of a morning is nothing to lose. This is no better than picking of pockets. I beseech you to banish this game from among you.

The other day the queen missed going to mass, and lost twenty thousand crowns in one morning. The King said to her, 'Let us calculate, madam, how much this is a year.' And M. de Montausier asked her the next day, if she intended staying away from mass for the hoca again; upon which she was in a great passion. I have heard these stories from persons who have come from Versailles, and who collect them for me.

Hoca (also known as basset) and lansquenet were games in which cards were dealt to banker and players, and then further cards were dealt face-up from the top of the deck (or, on unscrupulous occasion, the bottom); if they matched the cards held by a player, the bank would win; if they matched the banker's, players would win. There was opportunity for greater skill in other games, such as reversis, a forerunner of hearts, in which players tried to avoid the winning of tricks, and brelan, a forerunner of poker, in which there was opportunity for bluffing.

✳

The journals of Philippe de Courcillon, the Marquis de Dangeau, are much more reserved than those of his contemporaries and barely mention gambling at all.

Saint-Simon described Dangeau in later life as,

a harmless sort of personage… He was a tall man, very well made, grown stout with age, having an always pleasant face, which gave a promise, and kept it, of an insipidity that turned one's stomach… He was soft, obliging, flattering, with the air and tone and manners of society. People liked him because nothing ever escaped him against any one; he was kindly, accommodating, reliable in his dealings, a very honest man, obliging, honourable, but otherwise so flat, so insipid, such an admirer of nothings – provided such nothings related to the King or to persons in place and favour – so grovelling an adulator of the same, and, after his rise, so puffed-up with pride and silliness and so occupied with exhibiting and making the most of his pretended distinctions, that no one could keep himself from laughing at him.

Dangeau was born in 1638 to a Huguenot family in Brittany. He converted to Catholicism at an early age, and joined the army, where he distinguished himself in Flanders in campaigns against the Spanish. He served as a mercenary under the king of Spain before returning to France, where he set up a military academy, and became a colonel in the King's Regiment. Distinguishing himself again in battle, he resigned his commission and became a courtier and diplomat, applying himself to the task of accumulating titles and honours and money. He was appointed governor of Touraine, chevalier d'honneur to the dauphinesses of Bavaria and Savoy, counsellor of state, knight of the order of the Saint-Esprit, grand master of the royal and military orders of Notre Dame de Mont Carmel and of Saint Lazare de Jerusalem.

He was elected a member of the French Academy, even though he never published a book. Every Wednesday, the marquis and his brother, the Abbé de Dangeau, hosted a 'select party of literary men'.

Dangeau married twice, to Françoise Morin, sister of the Maréchale d'Estrées, and, after her death, to Sophia Maria Wilhelmina von Löwenstein-Wertheim-Rochefort, a countess of the Palatine family. Saint-Simon laughs at him for being so proud of his second marriage.

Dangeau was intoxicated by his success in pushing himself on, and when he reflected on the greatness of his wife's near relations his feet hardly touched the ground... where little credit is given to integrity and virtue, Dangeau always preserved a fair and unblemished reputation; but his conversation and his manners in general were less those of a nobleman of high birth and fashion, than of a man officious and obliging.

Saint-Simon's much livelier memoir of court life relied on Dangeau as a primary source. Whereas Saint-Simon's is caustic, his harsh eye dissolving behaviour into its motivations of cupidity and ambition and greed, or just pure silliness, Dangeau's account is a record of who went where, of the king's hunting expeditions, military manoeuvres, which foreign dignitary is visiting court, problems with Spain, royal pensions awarded, the laxity of morals at the English court of Charles II, performances of operas and comedies, costumes worn at Versailles balls. He records the deaths of notables, as well as some unfortunate accidents befalling the younger, wilder members of the court,

The Prince de Conti fell head foremost into the canal of Chantilly, but he soon came up again; he had swallowed a few mouthfuls of water, and was dragged out by the hair; two hours afterwards, he went to call on Monseigneur, and feels no inconvenience from his accident.

His accounts feature the same core group of eminences, the King and Queen, Madame de Maintenon, Monsieur and Madame, Monseigneur (the King's son), the Prince de Conti (who later would have become king of Poland if he had only got there in time, and died from a combination of gout and syphilis) and Princess de Conti, the Duc de Bourgogne (the King's grandson) and Duchess de Bourgogne – but, he barely writes about gambling. Dangeau does mention a fancy dress party in which the Duchess de Bourgogne danced as a deck of cards, and, in passing, that the queen hosted a game that lasted till three in the morning, or that the King had desired the 'grand game of reversis' to be played in his drawing room at Versailles. He won't say who was the big winner in the games. In July 1688, he reports that the Duc de Vendôme and the Prince de Conti fell out over an escalating argument at the card table, the matter seeming to be heading towards a duel before the King made an intercession.

And in May 1700, he writes,

> The Duc de Bourgogne recently asked the King for money. The King gave him more than he has requested and, as he gave him the money, told him how happy he was that he had approached him directly without asking anyone else to talk to him first. The King insisted that he always do that, and that he gamble without worrying as there would always be money and because losing should be of no importance to people like them.

Despite the description by Madame quoted at the beginning of this chapter, of the howling, blaspheming players, a seeming affectlessness was the correct way to gamble. Because it was only 'tolerated', there was no legal way of recovering debts. It was the fulfilment of the gamer's word that was at stake, an aristocratic attitude to the world, their place in it, the figure they cut, not an obligation to the creditor. One plays for the sport of it. Winning or losing, the gambler shrugs it off with equal equanimity – because they can afford to (*there would always be money*), because they are bred to (*people like them…*).

One paid one's gambling debts because one chose to, out of free will, and to show that one could. A dressmaker's children might go hungry because of unpaid bills, but a gambling loss, even to one of the many cheats in the court, was a debt of honour to be settled immediately.

The satirist Jean de La Bruyère wrote,

Nothing brings a man sooner into fashion and renders him of greater importance than gambling... I should like to see any polished, lively, witty gentleman, even if he were Catullus himself or his disciple, dare to compare himself with a man who loses eight hundred pistoles at a sitting.

La Bruyère was mocking a new era, where virtue and wit were proved at drawing room card tables rather than on the battlefield or in the service of poetry, as if there was something heroic about the gambler choosing to submit entirely to fortune. By displaying an indifference to mercantile values, the gambler was becoming an exemplary figure.

This was the world that Dangeau flourished in. One of the ways he had trained himself to do so was to make sure he would be underestimated by the people around him. Saint-Simon writes,

Dangeau had a mind below mediocrity, very frivolous, very incapable in every way, taking readily the shadow for the substance, living on gas, and perfectly contented with it. All his capacities went solely to behave himself properly, to hurt no one, to acquire, preserve, and enjoy a sort of consideration; without ever perceiving that, beginning from the King down, his silliness and conceit diverted the company, and that traps were laid for him in those directions, into which he tumbled headlong.

Even at moments of his demonstrated excellence, Saint-Simon barely allows the marquis any credit,

> With little wit, but what he had of the great world, the result of being always in good society, he allowed himself at times to scribble verses. The King had a fancy at one time for *bout-rimés* [a poetic game, in which unlikely words were given as a challenge to the player who had to make a rhyming sonnet out of them in the sequence they were given]. Dangeau was ardently desirous of a lodging at Versailles, when lodgings were scarce, in the early times when the King went to live there. One day, when he was playing at cards with Mme de Montespan, Dangeau sighed pathetically in speaking to someone of this desire, but loud enough for the King and Mme de Montespan to overhear him. They did so, and diverted themselves accordingly; then, finding it very amusing to keep Dangeau on the grill, they invented the strangest rhymes they could imagine and gave them to him, feeling quite sure he could do nothing with them, but promising him a lodging if he managed to compose them without leaving the game, and before it ended. It turned out that the parties duped were the King and Mme de Montespan. The muses favoured Dangeau; he won his lodging and received it immediately.

The muses favoured Dangeau... as if he was the passive recipient of the fortune of inspiration. As if he hadn't trained for this moment, at those weekly salons with his brother the abbé and other literary men. As if his 'pathetic' sighing hadn't been a perfectly judged display of seeming weakness that demanded to be tested, and rewarded, if he should pass the impossible test. (All poker players know the basic deception of 'weakness equals strength'; in a game of incomplete information, such as poker, such as life at the French court, sending misleading signals will generally pay off.)

According to Saint-Simon,

He had no means, or very little; he applied himself to learn perfectly all the games that were played in those days – piquet, bete, hombre, great and little prime, hoca, reversis, brelan – and to study the combinations of games and cards, until he possessed them so thoroughly as scarcely ever to be mistaken even at lansquenet and basset, judging them accurately and staking on those he believed would win. Such knowledge won him much; and his gains put him in the way of introducing himself into good houses and, little by little, at Court. Prompt and excellent at cards, where, no matter how great his gains (and they certainly were the basis and the means of his fortune), he was never suspected, and his reputation was always clean and intact. The necessity of finding heavy players for the game of the King and Mme de Montespan admitted him to their table; and it was of him, when fully initiated, that Mme de Montespan said merrily that no one could help liking him and laughing at him; which was perfectly true.

Madame de Sévigné, writing in 1676,

I was on Saturday at Versailles with the Villars. You know the ceremony of attending on the queen at her toilet, at mass, and at dinner; but there is now no necessity of being stifled with the heat, and with the crowd, while their majesties dine: for at three, the King and Queen, Monsieur, Madame, Mademoiselle, the princes and princesses, Madame de Montespan, and her train, the courtiers, and the ladies, in short the whole court of France, retire to that fine apartment of the King's which you know. It is furnished with the utmost magnificence; they know not there what it is to be incommoded with heat; and pass from one room to another without being crowded. A game at reversis gives a form to the assembly, and fixes every thing. The King and Madame de Montespan keep a bank together. Monsieur, the Queen, and Madame de Soubize, Dangeau, and Langlée, with their companies, are at different tables. The baize is covered with a thousand louis-d'ors; they use no other counters.

I saw Dangeau play!, and could not help observing how awkward others appeared in comparison of him. He thinks of nothing but his game, though he scarcely seems to attend to it; he gains where others lose; takes every advantage; nothing escapes or distracts him; in short, his skill defies fortune. Thus, two hundred thousand francs in ten days, a hundred thousand crowns in a month, are added to his account-book under the head received.

In 1678, Dangeau commissioned the young deaf mathematician Joseph Sauveur (who later made important advances in the study of acoustics) to calculate the banker's advantage in basset. Sauveur went on to make equivalent analyses of the odds in quinquenova and lansquenet. The elegiast Fontenelle wrote that Sauveur converted the games 'to algebraic equations, where the players did not recognise them any more'.

Edward Gibbon wrote of Dangeau, 'He saw a system, relations, a sequence, where others discerned only the caprice of fortune.'

And Fontenelle, in his elegy,

He had a supreme feel for gambling… he had penetrated its entire algebra, that infinity of relations between numbers which reigns within its various games, as well as all the delicate and imperceptible combinations of which they are composed, and which are often intermeshed with such complexity that they resist even the most subtle analyses…

M. le Dangeau with his head for algebra and full of the art of combinations, had many advantages… He applied theories only he could understand and solved problems only he could pose. Nevertheless he wasn't sombre like many serious players but talked and amused everyone.

This isn't quite right: Dangeau arranged for Sauveur's analysis to be published. The information was out there: impeccable Dangeau made sure of that. It's like the modern casino publishing tables of odds, complacent in the knowledge that the roulette players will carry on losing their money while drawing their patterns into the unknowable future. Dangeau, with his perfect manners, his soldier's discipline, the lightness and ease of his company, a shark among whales and fish, was getting it quietly and making everyone feel happy about their losses, even at times, despite his passion for etiquette and protocol, allowing himself to appear ridiculous in the process. He gained position, property, status, fortune, titles; everyone owed him money.

Except for Claude de Langlée. His adversary. The one other at court who knew how to win at cards without cheating, and who was also skilfully, although much less quietly, exploiting the aristocratic tendency to lose.

In Mme de Sévigné's words,

You know Langlée; he is as insolent as possible: he was at play the other day with the count de Gramont, who, upon his taking too great liberties, said, 'M. de Langlée, keep these familiarities till you play with the King.'

Gramont was one of the more notorious, and successful, cheats at court. 'His supper was always served up with the greatest elegance, by the assistance of one or two servants, who were excellent caterers and good attendants, but understood cheating still better,' wrote his biographer. Gramont wasn't the only court cheat. The dispute that Dangeau mentions in 1688 between the Prince de Conti and the Duc de Vendôme started when the duke, no doubt accurately, accused the prince of cheating.

Gaming is a kind of pyramid, with the many small losers at the base, which tapers to the top, to the small number of big winners. Losers pass money up to the fewer players above them. Of those at court who were at the apex of the pyramid, there were very few, maybe just three, who didn't cheat. Everyone would do their best to make sure the King won. Saint-Simon, who likes nothing better than to look on his subjects with the coldest, most cynical eye, states that neither Langlée nor Dangeau were ever thought to cheat at cards. Like the King, they didn't need to, albeit for different reasons.

Mme de Sévigné wrote to her daughter,

Monsieur de Langlée has made Madame de Montespan a present of a robe of gold cloth, on a gold ground, with a double gold border embroidered and worked with gold, so that it makes the finest gold stuff ever imagined by the wit of man. It was contrived by fairies in secret, for no living being could have conceived anything so beautiful. The dress was tried on; it fitted to a hair. Immediately it was concluded that it must be a present from someone; but from whom? was the question. 'It is Langlée,' said the King. 'It must be Langlée,' said Madame de Montespan; 'nobody but Langlée could have thought of so magnificent a present – it is Langlée, it is Langlée!' Everybody exclaims, 'It is Langlée, it is Langlée!' The echoes repeat the sound. And I, my child, to be in the fashion, say, 'It is Langlée.'

Langlée's mother had been a maid to the queen mother. His father had been a footman who had somehow accumulated the beginning of a fortune. It is unclear how this fortune was raised, because Langlée left behind no journals or correspondence; there were no elegies or biographies written about him, but Langlée inherited a bankroll, which he used to good effect.

Saint-Simon's account of Langlée's rise is almost respectful,

Born of obscure parents, who had enriched themselves, he had early been introduced into the great world, and had devoted himself to play, gaining an immense fortune; but without being accused of the least unfairness…

Similarity of tastes attached Langlée to Monsieur, but he never lost sight of the King. At all the fêtes Langlée was present, he took part in the journeys, he was invited to Marly [the King's informal residence, as opposed to Versailles, his very formal one], was intimate with all the King's mistresses; then with all the daughters of the King, with whom indeed he was so familiar that he often spoke to them with the utmost freedom. He had become such a master of fashions and of fêtes that none of the latter were given, even by princes of the blood, except under his directions; and no houses were bought, built, furnished, or ornamented, without his taste being

153

consulted. There were no marriages of which the dresses and the presents were not chosen, or at least approved, by him. He was on intimate terms with the most distinguished people of the Court; and often took improper advantage of his position. To the daughters of the King and to a number of female friends he said horribly filthy things, and that too in their own houses, at St. Cloud [Monsieur's residence] or at Marly. He was often made a confidant in matters of gallantry, and continued to be made so all his life.

Writing that 'Similarity of tastes attached Langlée to Monsieur', the King's brother, is Saint-Simon's uncharacteristically discreet way of saying that Langlée's sexual orientation was towards men; he had *le goût Italien*, the 'Italian taste', which, like gambling, was not legal but, in certain cases, tolerated.

His official position was *maréchal des logis général des camps et armées du Roi*, the King's quartermaster. But if you want fun, a dress, a party, the authoritative word on style, you go to Langlée. He is the only one who can gamble as high as the King and as successfully as Dangeau, actually even more so. And unlike Dangeau, he isn't restricted by what is appropriate, because nothing about him is appropriate, he's the *homme de rien*, a 'man of nothing', low-born; he can accumulate real estate from the gaming table, just so long as he continues to entertain – or provoke – everyone around him.

It wasn't just on matters of style and taste that Langlée was relied upon. Despite Madame's protestations to her sister ('Happily for me I no longer like cards, for I am not rich enough to risk my whole fortune...'), Saint-Simon noted that her losses at the table had been piling up,

Madame la Duchesse, whose heavy tradesmen's debts the King had paid not long since, had not dared to speak of her gambling debts, also very heavy. They increased, and, entirely unable to pay them, she found herself in the greatest embarrassment. She feared, above all things, lest M. le Prince or M. le Duc should hear of this. In this extremity she addressed herself to Madame de Maintenon, laying bare the state of her finances, without the slightest disguise. Madame de Maintenon had pity on her situation, and arranged that the King should pay her debts, abstain from scolding her, and keep her secret. Thus, in a few weeks, Madame la Duchesse found herself free of debts, without anybody whom she feared having known even of their existence.

Langlée was entrusted with the payment and arrangement of these debts. He was a singular kind of man at the Court, and deserves a word. For he was a sure man, had nothing disagreeable about him, was obliging, always ready to serve others with his purse or his influence, and was on bad terms with no one.

Except Dangeau. Who only occasionally writes about him. Sometimes his journal makes mention of a game, at which the King and Monsieur and the Queen are joined by Langlée and Dangeau, because no one else but those two could play as high as the royals. When, in 1700, Langlée was appointed by the King to take care of the gambling debts of another of his family, his daughter-in-law, the Duchess de Bourgogne, Dangeau mentions the fact of it, and that Langlée is 'a man of integrity and great regularity' – and if it disappointed him that the King didn't entrust him with this task or with the settling of Madame's debts, he doesn't mention it. But Dangeau never speaks of his own emotions in his journals, that deliberately dull gazette of court events.

Perhaps Saint-Simon is right, and Dangeau was as insipid as his journals indicate. If he wants an apartment at Versailles he can win one by turning his hand to verses. If he wanted more money, another title, he would win it at the table. Maybe he had no inner life, because he chose not to need one. Or it was so carefully repressed or hidden.

Dangeau would have kept records, money staked, his return on investment. Mme de Sévigné mentions an account book. He probably kept notes on dangerous opponents, those few others who, as Fontenelle says, were 'serious' about gambling. When he met with the mathematician Sauveur, his protégé, there would have been notes of that too.

There are no journals for the years preceding 1684. Perhaps he hadn't started keeping them yet. I prefer to think that he hadn't quite perfected the necessary insipidness.

Maybe there was a *real* journal somewhere, a continuation of his earlier one, before he had perfected the tone, and learned to separate his gambler's notes from his courtier's ones, a text that might have surprised Saint-Simon…?

Mme de Sévigné writes of Dangeau filling the zoo at Clagny, because he wanted to compete with Langlée and be as good at making presents as him. He paid 2,000 écus for the fattest sows and the fullest cows and the largest sheep and the best goslings and the most passionate turtledoves.

There was one public dispute between the two great gambling sharks of the Versailles court, when Dangeau's perfect manners failed him, as did Langlée's perpetual good cheer, in an argument about money owed, perhaps even, although this is undocumented, over how it was won, and by whom.

In a 1672 letter, Mme de Sévigné tells her daughter that 'MM. Dangeau and Langlée had bad words, at the rue des Jacobins, on a payment of gambling money… They were accommodated; they are both wrong. The reproaches were violent and unpleasant for both.'

If I were writing this as a novel, a major part of it would cover the period of the 'accommodation', the 'violent and unpleasant' reproaches between the two men, before they resumed their careers profiting on the probabilistic ignorance of the gamblers at Versailles.

Dangeau doesn't mention if he was offended by Langlée's manners. Or if he was envious of Langlée's accumulation of property and income, even from Dangeau. It is only that extravagant zoo that indicates his envy towards his rival. In the French national archives there's the agreement made between them in 1682 of a repayment schedule on a debt owed by Dangeau to Langlée. In the following year, there's a second document, a *quittance* declaring that half the debt had been settled.

The two great sharks of the Sun King's court then seemed to stay out of each other's way, as in the account by Madame de Sévigné of the entourages of Dangeau and Langlée gambling at separate tables. They would have taken it in turns to feed on the fish and the whales, the ones who relied on luck and notions of fortune rather than probability and odds. Langlée must also have been profiting from Sauveur's calculations, or similar ones, and those of his predecessors. There's a line that goes back to the Italian mathematician Gerolamo Cardano of attempts to calculate the odds at games of chance – and maybe that was the origin of Langlée's father's fortune too: the route for the footman to begin to establish his stake in the world.

*

Walter Benjamin wrote that betting 'is a device for giving events the character of a shock, detaching them from the context of experience'.

Last summer I was playing in a poker tournament in Las Vegas. At my table was the full range of poker types – the grizzled old grinder, grey-haired, rock-solid, who only played premium hands, and who kept himself amused by telling recent hospital stories and poker anecdotes from the old days, when the game was still a somewhat shady pursuit; the middle-class teachers and lawyers and accountants, who have taken to poker as a congenial hobby, a way to spend time without having to think about mortgages and family and work; the gamblers, of all classes and shapes, who are driven by the action, the movement of chips and luck and fortune; and the younger ones, who have learned the game online and are making a living at it, exploiting the older players' limitations, their lack of maths and theory and heart, and who by the time they're old enough to be allowed inside a Vegas casino have already clocked up more hands than the old-timers have played in their lifetimes. They often play in an aggressive, game-theory-optimal style that many of the old-timers neither understand nor appreciate.

In this case, the kid was a Brazilian. He got himself involved in a big hand against the old grinder. Betting went back and forth. The grinder only played big hands; the Brazilian could play just about anything. When the Brazilian kid reraised the old-timer, again, after a second king had hit on fourth street, the old-timer stared him down. The kid sat impassively, chin in hands, waiting for the older man to make his decision. Eventually the old-timer threw his cards away and patted the table in the gesture that signifies, *Well played, good hand*. His expression changed to irritated disbelief when the kid showed his cards. The kid had nothing, he was bluffing, and as he dragged in the chips he made a little explosion of joy. 'I love poker! I love it! The heart, goes ba-bump-de-bump-de-bump-de-BUMP. *BOOM!*'

What is one to do with those emotions? Where does the energy go, if you need to suppress it, in the interest of making your opponent fold, or for them to feel good about pushing their money your way?

In that moment when the Brazilian boy was being stared down there had been nowhere for him to store the adrenaline, the energy, the stress, that was rushing inside of him, so it raced through his bloodstream and made his heart feel as if it was about to burst. I suspect Dangeau and Langlée would only have felt that when they were going up against each other, because in other company even if they lost *this* hand, they would win it all back and more in the subsequent ones, because they were playing by principles of probability that their opponents didn't care to understand.

They were no longer submitting to luck; luck was submitting to them – they were being released from hope and fear, to something closer to certainty.

'Chance is *my* chance,' wrote Ella Lyman Cabot. Luck was *their* luck. And I'm getting closer to finding a definition of luck, something to whisper to Mrs Bevan to help her at her dinner table with the dying philosopher Wittgenstein.

Luck is the operations of chance taken personally.

CHAPTER 8

The Slopes of Vesuvius

Let us beware of saying that there are laws in
nature. There are only necessities: there is no one to
command, no one to obey, no one to transgress...
For believe me! – the secret of realising the greatest
fruitfulness and the greatest enjoyment of existence is:
to *live dangerously*! Build your cities on the slopes of
Vesuvius! Send your ships out into uncharted seas!

Friedrich Nietzsche, *The Gay Science*

In the wild love of chance, everything can be at stake.

Georges Bataille, *Guilty*

net total –£190

In 2004, Ashley Revell sold everything he owned, and with
the ensuing cash travelled from Kent to Las Vegas, where he
converted the money to chips, and stood in front of a roulette
table at the Plaza Casino, about to make a single $135,300 bet.

You can find footage of the event on YouTube, uploaded
by Revell himself from the TV programme that was following
him. The pit boss asks him, delivering his scripted line in a
rather stilted tone, 'Are you sure you want to make this bet
with your life savings?' Revell is thirty-two, blond-haired,

wearing a rented tuxedo and an open-necked shirt. He's playing the part of a high roller, but he's truly anxious, at the edge of something, not just acting it for the camera. There's a velvet rope behind which are family and friends along with the sorts of people who like to whoop it up in Vegas. One young woman appears to be wearing a cocktail on her head.

He knew he was going to make a 50:50 bet; he didn't know until the last moment whether he was going to bet on red or black.

'What colour then?' Revell asks the crowd, with an agitated laugh. There may be an element of stagecraft here – the tightrope walker almost stumbles, the juggler pretends to drop the knife – but as the croupier spins the wheel, Revell suddenly seems sure of something, 'Red,' he says, and moves all his chips into the corresponding box. The wheel turns, the ball rolls, one woman's hands are flapping in the air, she seems to be in some kind of breakdown; the woman with the cocktail headdress is smiling; Revell is standing still, mouth clenched, staring at the wheel – the footage that you can see on YouTube is slowed down for the final moments of the ball's journey, but that doesn't seem so cheesy at all, this must be what it feels like, to be caught in the moment of pure chance, pure opportunity, to be face to face with your luck. Revell takes a half-step forward, he crouches and jumps and punches the air as the ball finds its slot, and rests, at number 7, red.

He cashed in his winnings and returned home, where he resumed his life with twice as much money as he had before (minus US gambling tax). He bought a motorcycle and treated himself to a trip around Europe, where he happened to meet the woman who became his wife. He didn't play roulette again.

*

The Money had been calling. Where was I? When was I coming to LA? Were we still going to drive across the desert to Vegas? He had some of his own theories that he wanted to put to the test. How was my gambling coming along?

It was like talking to my father. You mean poker? I said. All right, poker, he said, how's the poker? I haven't been doing any of that, I told him. I've been taking a little sabbatical, concentrating on writing the book.

There was something about being in Dangeau's and Langlée's world that had made me ashamed to be playing low- and micro-stakes poker online. Jack 'Treetops' Straus once said, 'I wouldn't pay a ten-year-old kid a dime an hour to sit in a low-stakes game and wait for the nuts. If there's no risk in losing, there's no high in winning.'

'Nazi Gold,' the Money said.

'The Luck Book,' I said.

In the Samuel Johnson No-man-but-a-blockhead-ever-wrote-except-for-money tradition, the Money thinks I should be writing a pot-boiler, not these little literary novels and some all-encompassing Casaubon heart-sinker.

'How's it going?' he said.

'Slowly,' I said.

In the time that I've been writing this book, both my children have left school. My son has dropped out of one university and graduated from another. I'm over three years late in delivery. The project has gone through five editors at three separate publishers, and three different agents have represented it. The imprint that initially commissioned it has long since folded. The agent who negotiated its most recent contract is dead.

The Money's probably right: that's how it works, you make something to satisfy an appetite that you know exists because you've seen it rewarded in a marketplace (*Nazis! Gold!*), you get paid, your children eat, you buy a new shirt. The Money has made a lot of money writing and producing American network TV shows, heaps of the stuff that he has to keep spending, on real estate, on jewellery, because otherwise he falls foul of some accountancy tax retribution applied against the mega rich. He lives in Los Angeles, which is built on earthquake faultlines and liquefaction zones.

*

The truth was, I had been gambling. Sitting at my desk, in a more or less unvarying routine, I'd been reading German philosophers and Spanish aphorists, and I'd been betting on the stock market, hoping to understand and maybe channel the volcanic extremity of Friedrich Nietzsche by tempering it with the prudence of Baltasar Gracián. The only significant variation to my routine came from eating food suggested by other people's shopping lists.

Over all things stand the heaven accident, the heaven inno-cence, the heaven chance, the heaven prankishness.

'By chance' – that is the most ancient nobility of the world, and this I restored to all things: I delivered them from their bondage under purpose.

Nietzsche's word for 'chance' here, in *Thus Spoke Zarathustra*, is *Zuffal*, which is close to the English, or the Spanish *casualidad* and Italian *caso*, with their sense, deriving from the Latin *cadere*, of how things might fall. Because German has no word specifically for luck, classicists like Nietzsche turn to ancient Greek for terminology.

In a note from 1880, Nietzsche uses the Greek *ananke*, which is usually translated as 'necessity', to refer to 'the realm where everything proceeds arbitrarily (accidentally), where it is not the case that from every cause its effect must follow'.

Generally, 'necessity' is taken to mean what is impelled, inevitable, what unavoidably has to happen. Nietzsche's necessity is accidental, coincidental, contingent. The past is finished, the future is unknown.

Heaven is a dance floor for divine accidents... a divine table for divine dice and dice players. How did rationality arrive in the world? Irrationally, as might be expected: by a chance accident...

As creator, guesser of riddles, and redeemer of chance, I taught them to create the future and to redeem with their creation all that has been.

The philosopher Deleuze, allying himself with Nietzsche, combining the two of them into a presumptuous *we*, writes, 'When we affirm that chance decides, we are not thereby abolishing necessity, and vice versa. For necessarily there is one true chance combination: the game-changing roll. To affirm this roll as necessary is to affirm that all things come to be according to chance necessity... all such combinations in life, all events, are innocent chance necessities.'

Nietzsche isn't advising us to be passive in the face of this. As Dostoevsky knew, as Ashley Revell knows, the 'game-changing roll' doesn't just happen to us. And nor does luck exist outside of time. As Nietzsche writes, in aphorism 274 of *Beyond Good and Evil*,

The problem of those who wait – It takes a lot of luck [*Glücksfalle*] and all sorts of unpredictable factors for a higher person, in whom the solution to a problem lies asleep, to begin to act (or 'break forth', as we might say) at the right time. On the average, it does not happen, and in all corners of the Earth people are sitting and waiting, hardly knowing the extent of their waiting, and knowing even less that they are waiting in vain. Sometimes, too, the wake-up call – that chance event that gives them 'permission' to act – comes too late, when their best youth and strength for action have already been used up by sitting still. And how many a one, upon 'springing up', found to their dismay that their limbs were asleep and their spirit already too heavy! 'It is too late,' they said, no longer believing in themself and forever after useless.

Could it be that a 'Rafael without hands', taking the phrase in its broadest sense, is not the exception in the realm of genius, but rather the rule? Perhaps it is not genius that is so rare, but rather the five hundred hands it requires in order to tyrannise χαιρός [*kairos*], in order to seize chance by the head!

The Greek concept of '*kairos*' tends to be translated into English as 'occasion' or 'timeliness' or 'right moment'. This follows its Latin adoption as *occasio*, but that hardly does justice to the concept. The Latin removed some of the contingency, and urgency (which were transferred to the figure of Fortuna).

The god Kairos was Zeus's youngest son who was customarily pictured with winged feet, and carrying the balancing scales of fortune along with a razor, and wearing a distinctive hairstyle of a loose forelock in front and bald pate at the back.

A bronze statue of Kairos sculpted by Lysippos was famous in the Greek world. Carved into it was an epigram by Poseidippos,

Who are you? *Kairos who subdues all things.*

And why do you stand on tip-toe? *I am always running.*

And why do you have a pair of wings on your feet? *I fly with the wind.*

And why do you hold a razor in your right hand? *As a sign to men that I am sharper than any sharp edge.*

And why does your hair hang over your face? *For him who meets me to take me by the forelock.*

And why, in Heaven's name, is the back of your head bald? *Because none whom I have once raced by on my winged feet will now, though he wishes it sore, take hold of me from behind.*

Why did the artist fashion you? *For your sake, stranger, and he set me up in the porch as a lesson.*

The forelock is there to enable us to grab at opportunity ('to seize chance by the head!' as Nietzsche wrote), and he's bald at the back because once opportunity has been missed the moment is irretrievable.

Homer's *Kairos* was the point at which to aim an arrow, the most vulnerable part of the target, the soft spot; this could be a gap in a fortification or an unprotected part of the human body, which is where we get our word 'temple', from the Latin *tempus* for 'time', to designate the weakest point in the skull. Similarly, 'opportunity' derives from the Latin *porta*, an opening.

There are three primary overlapping meanings in the concept of kairos. It is timeliness; it is a turning point, a critical moment that calls for decisive thought and action; and it is individual opportunity.

Greek rhetoricians built structures of persuasion on the subject. Greek physicians wrote about kairos as the crucial moment in treating a disease. Hippocrates' aphorism 'Life is short, art is long' continues with 'Kairos is fleeting'.

A German lyric poetry tradition, which Nietzsche would have been familiar with, developed from Pindar's victory odes and Cicero's writing about *occasio*. It explored the notion of the 'privileged moment', the time for acting or not acting, and the 'fulfilled moment', in which time is redeemed. Goethe wrote about this in *Faust* ('The clock may stop, its hands fall still, / And time be over then for me!') and Schiller in *Don Carlos*, 'One must take advantage of a moment like this, which comes but *once*.'

These are the moments when we act, as Plato said, 'in accordance with chance out of necessity'.

Messianic thought is kairotic, the redeemer coming in the fullness of time. Every process or tradition which has a notion of 'the right time' or 'timeliness' or 'crisis' is kairotic. But let's strip away teleology and notions of redemption: for us, it's the apprehension of the opportune moment, when we might bring our experience and flair to bear upon the event, to turn it our way.

Luck does not exist outside of time. Opportunities have to be recognised. Too many slip by without us even knowing that they were there. Kairos went past when we were looking the other way.

The concept of kairos informed Machiavelli's ideas of prudence, fortune and opportunity. Before writing *The Prince*, he composed a poetic, similarly dialogic, version of Poseidippos's epigram. Although, following his Latin models, Machiavelli's version of opportunity was now a woman.

And she had a companion, named Penitence, because, as Machiavelli has Kairos say, 'whoever does not know how to capture me will get her'.

*

I went to see an investment fund manager. I told him I was deepening my study of chance and luck. I wanted it to be a practical study.

'You mean risk,' he said.

'Maybe. I'm not sure. Yes,' I said.

'There's an expression in investing, *Early is as bad as wrong.*'

'Kairos,' I said.

He seemed to mistake that as a cough.

'You'll be fine,' he said, 'as long as you understand probability. Do you understand probability?'

I illustrated to him what I understand of probability.

If I need one card of a certain suit to complete a flush and there is one more card to come, then I have a 1:4 chance of making my hand (there are four suits in the deck). If there is already $100 in the pot and it is going to cost me $20 to see the river, that's giving me the price of 6-to-1 on a 4-to-1 chance, which is irresistible (there is now $120 in the middle: that $100 plus my opponent's $20 bet, for which I have to spend $20 to try to win). But maybe my opponent, 'the villain' as they say in poker problems, is trying to make sure that hands such as mine, which are guaranteed to beat his, don't get the correct odds to stay in the hand. He makes it $50, not $20. Now I, 'the hero', have a price of 3-to-1 on the same 4-to-1 chance. These are not good odds.

Except. There is the notion of *implied odds*. We each began this hand with a thousand in front of us, and I am confident that the villain is not backing down no matter what. The potential benefit is not limited to the $120 in the middle, it's everything that is on the table. That changes the equation. My implied odds now are better than 50-to-1.

'That's not quite what I meant,' Roddy the trader said.

He asked me a variation on a classic game theory question. There are ten doors; behind one of them is freedom, behind the other nine is death. You're the prisoner and you have to choose, without any further information, one of the doors to open. You select door two, and all the other doors are now barred to you. Eight of them are investigated. Luckily, these are all death doors, and you've escaped them. Two doors are left, the one you originally selected, and one other. You know that behind one is freedom, behind the other is death. Now you're given the opportunity to change your mind. Do you?

As it happened, I'd had a conversation with someone a couple of days before about this, so I knew the correct answer.

Originally the prisoner had a 10-to-1 chance of guessing correctly – one door in ten would lead to freedom. Now it's 2-to-1. With probability as a guide, you have to ignore your loyalty and any sentimentality or superstition about your original choice. You must change your mind, take the option of the 'better price' of the alternative door.

By saying yes to his question, I gave Roddy the mistaken apprehension that I'm more advanced in my understanding of probability than I am.

'That's basically all you need to know,' he said. 'If you understand that, you can make money out of the markets. Drink?'

We drank. I asked him how he had made his money.

'By understanding probability better than the next person. And shorting. I seem to be good at betting against companies. I keep trying to bet on them, to find a little company that's going to grow, but generally I'm much better at finding companies that are about to tank. You should try some spread betting. See how you go with that.'

So this is what I've been doing, which I haven't told the Money about.

To focus my mind more closely on the subject, I've opened an account to make spread bets on financial markets. This might be a purer form of betting than most, or just more decadent: to bet the 'spread' is to speculate whether the price of a commodity, which one doesn't actually own, is going to rise or fall.

The first two bets I placed were to 'sell' the London Stock Exchange and Wall Street. It seemed obvious: the chart for the London Stock Exchange had shown a steady fall over the morning before levelling – I assumed as the traders waited for Wall Street to open. The US government was in turmoil; it seemed obvious that trading would go down. I bet 25p a point on the London price and £1 a point on the NY price. I put 'stop-losses' and 'stop profits' in place, so my bets would automatically be cashed out, my positions would be closed, if they reached a specified low or high.

Instantly, both prices went up, putting me into the negative red. I watched the numbers rise at the bottom of the screen; and then the momentum shifted, the prices were going down. I was in green profit. I had made nearly £1 on London, £5 on Wall Street. I could take my profits now, dream of having bet £100 a point instead of one: this was easy; and then the prices climbed again. Now I was £2 down in London, £8 down in New York. Even though the tendency on the NY chart was to slump and then rise higher, I was sure that I wouldn't reach my stop-loss (of £20) and that I would just have to hold my nerve and wait for the market proper to open.

170

The red and green spikes on the City Index charts exerted their own horrible fascination.

The NY trade reached my stop-loss, so it automatically closed and I bought it again. And lost again – and still the market hadn't even opened yet. This was like the poker player Stu Ungar taking $50,000 with him to bet on a golf game he was about to play, and losing it all on proposition bets before the players had reached the first tee.

Except it wasn't like that at all. Stu Ungar lived on the slopes of Vesuvius, fuelling himself and ruining himself with cocaine and gambling, and died there, at the age of forty-five. Or, if not Vesuvius, a hotel room in Las Vegas, burnt out and hollowed out, a year after he had come back from destitution to win the Main Event for a third time. I wasn't even on a foothill.

I bought Germany; I liked what I could intuit from its graph, and I supposed that it would be going the opposite way to the UK and the US. I closed that down at a £12 loss.

When Wall Street finally opened, it plummeted. My profit was £20, I waited for the red line on the screen to go down further, but then it rallied so my profit was only £6.50. I hated seeing the market rally. But then it fell again, a sharp decline. I sold my position when I'd made a profit of £38.50. I then closed the London Stock Exchange for a £3 profit.

I was £22 up on my trades for the day. Pleasingly, both Wall Street and London quickly recovered from their early positions.

I logged out and logged back in again to make a couple of pounds betting on the euro's strength increasing against the dollar, but I closed because the changes weren't volatile enough.

The action was different to what I was used to. It was addictive and it was stressful, and it was thrilling. Gambling is usually a response to the odds or perceived odds on a finite event – a horse race, a lottery draw, the turn of a card – here, the action was perpetual, until the stop-loss was reached, or the nervous punter clicked the 'close' button.

The following day, determined to be braver, I held my position on the London Stock Exchange, even when I had made twenty times my original stake. It then rose and rose and rose and I lost twenty times my stake, and closed my position, to wait for what I assumed would be the inevitable downturn to sell it again.

I hadn't read the financial papers, I didn't subscribe to any financial wizard's newsletter. I wasn't being guided by my superior understanding of probability against all the villains out there who were on the other side of my bets. The green and red lines of profit and loss on the graph looked like mountains. I was experiencing a constant rush of adrenaline and anxiety about my positions, as they're called, as if this was ballet, or yoga. I felt as if I was living a perpetual Nietzschean moment. It's been pointed out before, that with chaos theory, which describes a world that has to adapt itself to a condition of never-ending novelty at the edge of an abyss, the scientific world view is beginning to catch up with Nietzsche's inward vision (outwardly, he was protected by thick green lenses from the sunlight that was so violent to his susceptible physical system).

But even Nietzsche would acknowledge that there's a role for prudence here, a set of principles to guide our hand as we reach out to grab Kairos by the forelock before his wings pass us by and the bald back of his head gliding past is all we touch, leaving us, as Machiavelli would say, with Penitence.

*

Baltasar Gracián y Morales published *The Oracle: A Manual of the Art of Prudence* in 1647 to official disapproval. Nietzsche, who encountered the *Oracle* in Schopenhauer's German translation, wrote of it, 'Europe has never produced anything finer or more complicated in matters of moral subtlety.'

The *Oracle* is a collection of three hundred maxims for practical success, in condensed, sometimes paradoxical form, written by a seventeenth-century Jesuit, who was probably descended from Jewish converts. Its ideas about fortune have been packaged in the modern day as self-help recipes for success.

Maxim 17
Vary your way of behaving.

Do not let it always be the same; so that you may puzzle anybody who is watching you, and more especially if he is competing with you. Do not always act openly, for your rivals will recognise this uniformity of behaviour and will forestall, and even frustrate, your actions. It is easy to shoot a bird on the wing if it flies straight; not so, one that twists and turns. Neither should you always behave disingenuously, for people will see through the trick the second time you try it on. Malice is on the watch; great skill is required to circumvent it; the sharper never plays the card his opponent is expecting, and still less the one which the latter wishes him to play.

A contemporary poker player would recognise this advice as 'Balance Your Range', which is a tenet of modern game-theory-optimal strategy. Sometimes with aces in late position you might just call, rather than make the obvious raise. Sometimes with a marginally playable hand in early position you might four-bet (bet, and then if someone raises you, bet again), as if your hand is much stronger than it actually is.

You might even use a random procedure to decide when to do this. If a call and a raise are equally good options, then you might toss a coin or its equivalent – looking at a digital readout to see if it shows an odd or even number, say – to decide which to do; if you like to bluff a quarter of the time in one particular recurring situation, you might choose to do so when the second hand on a watch face is within the first quadrant.

I didn't know how this maxim would apply to spread-betting on the stock market online, but I was beginning to feel as if I'd had enough of that. I had got lucky at the beginning. When I closed my eyes, shifting green and red lines spiked up and down. I was considering moving on.

38

To know how to abandon the game when their fortune is in.

A good retreat is as important as a spirited attack; it safeguards your achievements when they are adequate as well as when they are numerous. Continuous good luck is always open to suspicion; intermittent good fortune is sager, and let there be an element of the bitter-sweet about it even while you are enjoying it. The more your strokes of good luck crowd upon one another, the greater the risk you run of their slipping and bringing down the entire heap. The transience of a favour may sometimes be counterbalanced by its lavishness. Fortune soon tires of carrying one too long upon her shoulders.

Thank you, Gracián. I'll close the account now.

After writing about bibliomancy, I had tried the *sortes Virgilianae*. I had asked the *Aeneid* if I am lucky and it took me to where Aeneas tells Dido the dream he had while asleep as the Greeks ransacked Troy, in which he had been visited by Hector, 'with ragged beard, with hair matted with blood, and bearing those many wounds he received around his native walls. I dreamed I myself wept, hailing him first and uttering words of grief.'

Which had been a disappointment, to have Hector the doomed hero presented as my emblem of luck. 'Bearing those many wounds', dream-Hector has come to warn Aeneas to flee the burning city of Troy. This dream apparition saves the hero. But all the same I wasn't sure that this was the sort of identification I wanted to have for luck.

I worried over what else the passage might tell me: to trust in dreams? That good advice – or the stirring of fortune – can come in unlikely places, and in hideous guises? That somehow, despite seeming misfortune, despite Trojan horses, and shipwrecks, and devastating love, I too might found Rome?

Consulting the *I Ching* with the same question, I cast the hexagram, 22 *Pi* ('Grace'), which had once alarmed Confucius. Asking if I was lucky, I was told, somewhat dispiritingly, that 'the fire, whose light illuminates the mountain and makes it pleasing, does not shine far'.

I tried to put these moments with the oracles out of my mind. Gracián was a preferable guide.

113
Plan for bad fortune while your fortune is good.

170
In all matters keep something in reserve.

Again, this is recognisable as wisdom from poker experience: the standard advice is never to expose more than 5 per cent of your bankroll at any time. If you have $100 to gamble with (and always have your gambling money separate from domestic accounts, from the household petty cash), then the maximum you should ever risk from that is five. Lose that, and you should play with slightly less the next time, a new 5 per cent. Failure of bankroll management as much as addictions to cocaine and gambling was the reason Stu Ungar went broke. It is the reason most players go broke.

In the absence of prudent management, you're relying on variance, on luck always swinging your way. A few weeks after French Eric went broke, he noticed something glinting from the pavement in Stockwell. It was a £100 chip from the Empire Casino, which, to the uninitiated, probably looked like play money. He picked it up and went to the cardroom that evening, and built his bankroll back up, promising himself – and lying to himself – that he would be more prudent next time.

172
Never compete with someone who has nothing to lose.

The struggle will be unequal. One of the contestants enters the fray unencumbered, for they have already lost everything, even their shame...

But, if I'm being Nietzschean, shouldn't I be the one who has reached the point where I have nothing further to lose, who is prepared to stake everything, who, even in my heart, might already have? Might I be using a notion of 'prudence' to cover what I suspect of myself, the limits of my courage?

182
Show everyone a bit of daring.

My life shrinks to this project. I'm writing about risk and jeopardy and volcanoes and I'm doing it in a well-insulated room, where I can pull the blind down against the sun because there's a glare on my laptop screen; and already I'm looking for excuses not to play, to pull back from the biggest gambles. Think of Ashley Revell. He staked everything, and in his holiday after returning from Vegas he 'happened' to find a wife. Maybe it was the experience of risking it all that enabled this to happen, that focused his mind and his senses so acutely to see opportunity where it arose, that otherwise, even on the same motorcycle, the same European adventure, she might have slipped past unnoticed.

But I am already married, and happily, or luckily, so. Maybe I should empty the shared domestic bank account, take the proceeds to Vegas, make that journey across the desert, and risk it all. That would *Show everyone a bit of daring.*

What would Dangeau and Langlée do? They probably had read the *Oracle*, or maybe they came to its conclusions on their own.

106
Don't flaunt your good fortune.

107
Don't look self-satisfied.

112
Win the good will of others.

118
Be known for your courtesy.

119
Don't make yourself disliked.

124
Make yourself wanted.

127
Ease and grace in everything.

226
Be sure to win people's favour.

295
Not a braggart but a doer.

89
Know yourself.

As Nietzsche wrote, adapting a line from Pindar, 'Become what you are, once you know what that is'.

274
Be charming.

Nietzsche was known for his courtesy, especially to older women. He had good manners, dressed carefully and appropriately. His health had been precarious from childhood, but before it collapsed, apart from some moments of agitation, he presented a composed, quiet appearance to the world. As he said, 'The gentlest and most reasonable of men can, if he wears a large moustache, sit as it were in its shade and feel safe there...'

Gracián doesn't mention facial hair in the *Oracle*. Pictures of him show a clean-shaven man under a three-cornered clerical hat.

Maybe I should grow a moustache? Or find a less shady place to feel safe in?

210
Know how to handle truth.

225
Know your major defect.

Poker is the revelation of character, and mine might lack something. Should I reopen the spread-betting account, and remove the stop bet? Before being permitted to open an account in the first place, I had to talk to an account handler, to convince him that I had enough money to lose, that I was not going to bring ruin on myself and bad publicity on his company.

One of my weaknesses as a poker player is contentment, almost an indolence, when I'm in a strong position. I tend to sit back, remove myself a little from the fray when, of course, I should be doing the opposite, using my large chip stack to bully smaller ones, to gain more.

The converse must be guarded against also, the splashing around of chips, throwing everything away because it costs too much psychic energy to hold on to it.

I'm beginning to understand why Ashley Revell went to Las Vegas. We have had a few brief conversations about arranging to meet, to talk further about his grand gamble. But each time we'd nearly had an arrangement, he'd pulled away from it.

What would Nietzsche do? Despite his weak constitution, the chronic headaches, his poor digestion, his propensity to vomit, to diarrhoea, I have no doubt that he would climb a mountain. It is hard to imagine him gambling on a laptop computer. Except, there is something masturbatory about the process, the solitude, the closed system. Freud diagnosed Dostoevsky's gambling as masturbatory. Wagner was sure that Nietzsche was a chronic masturbator. It was finding out about the correspondence on the subject between the composer and Nietzsche's doctor that drove the philosopher to make his final, furious break with his master. (So, yes, I can picture him as an online poker player, and a good one. While Dostoevsky would go repeatedly broke, Nietzsche's daring – if he could manage his bankroll – could take him to the top of the pile.)

Like all the best self-help books, or acts of prophecy, Baltasar's messages can be contradictory, sometimes drawing attention to this by reversing the advice in neighbouring aphorisms:

277
Display your gifts.

278
Don't draw attention to yourself.

216
Express yourself clearly.

253
Don't express your ideas too clearly.

Baltasar was a near contemporary of Thomas Bastard. In his earlier book *The Hero*, a response to Machiavelli's *The Prince*, Baltasar had sought to identify the qualities which characterise the ideal leader and achiever. The paramount quality is *despejo*, which contains Bastard's notions of virtue and wit, but is rather more dashing: 'Bravery, quickness, intellectual subtlety and wit... It is a mysterious formal quality, the highest quality of any quality... Being the soul of beauty, it is also the spirit of prudence; and being the breath of elegance, it is the very life of bravery...'

I am convinced by Baltasar that courage and prudence are necessary companions, that my success at luck will come as a result of finding both of them in myself, my inner *despejo*. Most of what Baltasar says is convincing, but there are times when the way he says it draws maybe too much attention, when he seems to be at similar pains to Thomas Bastard to display *ingenium* in his language. Borges wrote a grumpy poem about him, in which his disregard for his ornamentalist, mannerist subject is not entirely free from self-reproach,

Labyrinths, puns, emblems,
a cold, over-worked trinket,
poetry for this Jesuit
was reduced to stratagems.

There was no music in his soul; just a vain
herbarium of metaphors and deceits
and a veneration of conceits
and for human and superhuman, disdain.

That 'superhuman' refers to God, to angels, to any aspect of a manifold divine, but it is also an anachronistic reference to Nietzsche, who would have applauded the aphorist for his inconsistencies.

*

So, derived from Nietzsche, here's the new predicate: there is no external intention, there can be no generalised purpose. Even if we were not so restricted in our position and our thinking, we still would not see how it all connects, because it does not; there is no wise or merciful or malicious or indifferent mind behind it all. There are no necessary connections. It is all chance. Everywhere is risk.

And yes, we could be making our relatively safe place to be, huddling together for warmth and security, while growing our moustaches, but the more authentic response is to embrace this state of things. Develop *despejo* to help us find and seize kairos.

Thus Spoke Zarathustra begins with the prophet's first disciple, a tightrope walker, falling and dying. It is not just being on the slopes of Vesuvius that interests Nietzsche, it is the volcano itself. When he was still at school, the future philosopher began writing a play, *The Death of Empedocles*. The healer-philosopher Empedocles, probably to achieve divinity, chose his death; according to the poet Horace, he 'coolly leapt into burning Etna'.

This is why volcanoes and high buildings are so frightening. The French phrase for it is *l'appel du vide*, 'the call of the void'. I am not afraid that the volcano might erupt, but of what I might do to myself when I get there.

Walter Benjamin had this to say on gamblers,

the loser tends to indulge in a certain feeling of lightness, not to say relief... it is only at the last moment, when everything is pressing toward a conclusion, at the critical moment of danger (of missing his chance), that a gambler discovers the trick of finding his way around the table, of reading the table... Furthermore, one should note the factor of danger, which is the most important factor in gambling, alongside pleasure. ... The particular danger that threatens the gambler lies in the fateful category of arriving 'too late', of having 'missed the opportunity'... gambling generates by way of experiment the lightning-quick process of stimulation at the moment of danger, the marginal case in which presence of mind becomes divination – that is to say, one of the highest, rarest moments in life.

Ashley Revell could have made his same bet at a casino in London, or in a bookmaker's in Kent. Online gambling already existed in 2004, but there isn't anything particularly telegenic about watching a man looking at his computer screen. And it is less meaningful to do it at home. He crossed an ocean, travelled to an oasis in the desert. Vegas stands for something, a mythology that feeds the gambler's yearning as well as the resort's publicity. Luck is not static, and nor is it quite ever here; the hero makes the journey; the tightrope walker falls; you have to reach out to catch Kairos by the hair, but it's gone. Luck is the thing just outside the edge of oneself.

financial spread betting £147

CHAPTER 9

The Measurement of Uncertain Things

All our knowledge is based on relationships and
comparison, everything is therefore relation in the
Universe; and hence everything is subject to measure…

Georges-Louis Leclerc de Buffon, 'Essays on
Moral Arithmetic'

And indeed, where years were concerned, General
Yepanchin was, as they say, in the very prime of life,
fifty-six and not a day more, which is of course a
flourishing age, an age when *real* life truly begins.

Fyodor Dostoevsky, *The Idiot*

net total –£43

After closing my spread-betting account, I returned to poker.
And I started losing.

The temptation when one's luck is running especially badly
or especially well is to enlarge the space in which it can operate. Prudently, when we are playing an inferior opponent, we
should be reducing the scope of luck, to allow our superior
skill its chance to operate. In a tournament, we do not want
to be taking 50:50 flips, especially when the weaker opponent
has more chips than we do.

When we are 'in form', we expect those flips to go our way (and this 'expect' is really just a species of hope), so we take the bigger chances, promising ourselves that the risk is negligible, the reward is great, even if the numbers don't back that up. But when we are out of form – and I was out of form – we fear that even our good hands will find a way to lose, we will probably be rivered anyway, so we might as well surrender ourselves to the operations of luck, and to the experience of the loser's relief: the voluptuousness of pouring away money, the abdication of control, a yielding to something bigger, the sickly sweetness of surrender. Because, when you surrender hope, you become immune to fear.

I was losing quickly, and slowly, and then quickly again. I realised I had started playing very defensively, expecting the chanceful element of the game to work against me. So I changed my game and I was playing aggressively, surprising a new table with my style, because the older gentleman does not usually play in such a whirlwind way, before they saw that this was not selective aggression, it was the loosest of Loose Aggression; and all they had to do was to wait to pick me off with a premium hand as I fired off barrel after barrel, as my two pairs on the flop lost to a straight on the river, as my flush draws never came in. This wasn't *despejo*, it was *la imprudencia*.

Several times, I would walk down the staircase of the Vic away from the poker room, alarmed at these tendencies in myself. *Have a good night*, one of the casino staff would say, slightly disapproving as I put on my cyclist's reflective jacket and checked my lights, and I'd go out into the night, to cycle back to South London. If it was early enough, and it often was during this losing streak, Hyde Park would still be open, and I would cycle through that, and there was some kind of balm in the air, the edges of myself were soothed, a little. But something was still inside of me unabated, a kind of frenzy.

If the park was closed, I would cycle down Park Lane, take my chances with the traffic there, the limousines that pull suddenly away from the hotels on the left without indicating, the taxis and Uber drivers that waver and swerve in front of you, and it would require the closest of concentration to stay out of trouble; this was maybe the concentration that was lacking at the poker table, and there was something worrying in all this. I felt like the kind of gambler who chases losses by going up a level and then another. I was continuing to raise the stakes – *yes, I've lost my money, but what are you going to give me for my LIFE?!* – because my relationships with hope and fear had become jaded.

*

After making his fortune from the lottery in collaboration with Voltaire, the mathematician Charles Marie de La Condamine used some of the proceeds to fund expeditions to Turkey and South America. The French Geodesic Mission mapped the Amazon and measured the circumference of the Earth. Condamine also helped to standardise the metre as the unit of length, which was defined as one ten-millionth part of the distance from the North Pole to the Equator.

According to a legend that was written down by the historian Josephus, it was Cain, the first murderer, who devised weights and measures. Cast out after the killing of Abel, he founded a city (legend is unclear as to where the inhabitants of the city came from), where he introduced the means to parcel up – and profit from – God's bounty.

The French Enlightenment 'project' that Condamine was part of was a kind of Cainish fury to quantify the world. In the second half of the eighteenth century, naturalists, mathematicians, adventurers, explorers, collectors set out to measure everything and to catalogue creation and to ascribe to every thing a value. They compiled encyclopaedias and constructed taxonomies of animal species and chemical elements.

Georges-Louis Leclerc, comte de Buffon, was the author of the monumental *Natural History, General and Particular, with a Description of the King's Cabinet*, which was published in thirty-six volumes between 1749 and 1789 (and eight more after Buffon's death). But as well as external bodies and phenomena, the measuring was to be of internal events too. In 1777 he produced his 'Essays on Moral Arithmetic', in which he wrote,

> The measurement of uncertain things is my object here. I will try to give some rules to estimate likelihood ratios, degrees of probability, weights of testimonies, influence of risks, inconvenience of perils; and judge at the same time the real value of our fears and of our hopes.

The real value of our fears and of our hopes... This is a grander enterprise than measuring the length of the Amazon or the circumference of the globe. Buffon is setting out to devise, construct and then standardise a measure for our emotions. He starts with fear and hope, which he sees as opposite points on the same line,

> After having reflected on it, I have thought that of all the possible moral probabilities, the one that most affects man in general is the fear of death, and I felt from that time that any fear or any hope, whose probability would be equal to the one that produces the fear of death, can morally be taken as the unit to which one must relate the measure of

the other fears; and I relate to the same even the one of hopes, since there is no difference between hope and fear, other than from positive to negative; and the probabilities of both must be measured in the same way.

The scale, according to Buffon, is made up of gradations derived from the 'moral certitude' that we will live another day.

All fear or hope, whose probability equals that which produces the fear of death, in the moral realm may be taken as unity against which all other fears are to be measured.

The base unit he chooses to measure from is a previously healthy man of fifty-six years old. This is the age when, Buffon says, 'reason has attained its full maturity and experience all its force'. (Buffon was sixty-nine at the time when he was devising his moral arithmetic; I happen to be fifty-six writing this.)

After consulting actuarial tables, Buffon calculates that the probability that a previously healthy man of fifty-six will die within the next twenty-four hours is 1 in 10,189. (He was criticised for this calculation, which might be more accurately 1 in 100,000. Buffon's unperturbed response was, 'This difference, although very large, changes nothing of the main implications that I draw from my principle.') Any event whose probability is less likely than one chance in 10,189 (or one chance in 100,000) is one we can be reasonably and, therefore, morally, certain will not occur, and one to which we should accord no more real importance than we do to the possibility that we may die in the next day. 'From this I conclude that any equal or smaller probability must be regarded as zero.'

The prospect of any event for which the odds are equivalent or longer to me dying within the next twenty-four hours, 'should neither affect nor occupy our feelings or our minds for a single moment'. To put it another way, in the standard terms of probability, where 1 represents certainty and 0 impossibility, the chance of me dying suddenly within the next day is .0001 (or .00001), which becomes the operative zero point on the scale of moral probability.

But if death is the zero point on the scale of fear, what is at the opposite end, the '1' of hope?

What can we dare to hope for?

What is hope when we've got nothing? When we have a plenitude? What is the opposite of death?

Hope is faith – that the messiah will come, that God extends a hand, that the enemy soldier will shoot somebody else, that the lottery will reward me with a win. The intervention will come from outside, the change that I long for will be provided by a greater power. If I could only have *that* all my worries will be over, because luck and happiness have merged into a single identity again. The future will be like now except with that one thing changed. I, consequently, will be happy. I, consequently, will be new.

At either end of the scale, Hope (lottery win, eternal life) and Fear (my death) can have the same effect – eliminating agency and responsibility, making the individual subordinate, inactive, still, waiting to be obliterated by something stronger than self. It is in the middle where we must act with prudence and courage. At the poles of Buffon's moral scale, the individual is frozen in inaction. If hope or fear approach certainty, their terminal points, there is nothing to do, there is nothing that one can do: the messiah is on the way, my death is certain.

Kant called this state of utter dependency *Unmündigkeit*. It is the 'lack of resolve or courage to use understanding without guidance', and literally means 'unmouthingness', having no power to say.

Stephen Crane wrote, in a letter to a friend, 'Hope is the most vacuous emotion of mankind.' He had less than a year left to live when he wrote that.

How then might we make any plans? Should we not adopt the attitude of Jocasta in Sophocles' *Oedipus the King*?

> But why should man fear since chance [*Tyche*] is all in all for him, and he can clearly foreknow nothing? Best to live lightly, as one can, unthinkingly.

Buffon used probability theory to try to reason lottery players away from their vice. In England Daniel Defoe did a similar thing in his tract 'The Gamester'. But reason has nothing to do with it. That the game is so difficult to win and leads to rollovers enhances its attractiveness. The lottery sells hope, which Kant defined as the 'unexpected offering of the prospect of immeasurable good fortune'.

Barbara still hopes for a lottery win. Even though there is much more chance of a previously healthy fifty-six-year-old man, her next-door neighbour, say, dropping dead in the next twenty-four hours than of her winning a lottery.

For Buffon, and probabilists like him, only a fool would invest in a single lottery ticket when more than one hundred thousand other tickets are sold. Barbara would have to buy 2,258 tickets just to give herself the slightest chance. She does not intend to buy 2,258 tickets. She buys one for herself, and one for each of her grandchildren. Her chances are, in Buffon's scheme of things, zero.

*

Fear of death is intrinsic to modern probability theory. There were other contributions to its development, including the Christian literalist urge to verify episodes related in the Bible, and a legalistic reckoning of the likelihood of the evidence to determine an accused person's guilt.

Buffon's arguments against the lottery point back to the primary impetus, a prudent gambler's need to find a way of estimating chances.

Gerolamo Cardano's *Liber De Ludo Aleae* ('Book on Games of Chance') is the first probabilist text to be written. Cardano (1501–1576), based in Northern Italy, was, among other things, a doctor, a mathematician, a memoirist, an astrologer, and a gambler. He was unlucky in his astrology: casting a chart for King Edward VI of England, Cardano predicted serious illnesses at the age of twenty-three and thirty-four, marriage, a long reign. The sixteen-year-old king died a few weeks after the consultation, which brought ridicule to his astrologer's practice.

Cardano was unlucky in many things. He begins his autobiography with an account of some of his misfortunes, and he comes to a conclusion that is similar to Machiavelli's: that half of what we experience is out of our control, but we might be able to do something about the other half,

Let us therefore examine what this luck is and on what principle it depends; certainly it seems to me to be a disposition of affairs in accordance with or adverse to the will or plan of a man; so that no matter how you act, the matter turns out well or badly, or agrees with human plans or does not agree. Good fortune is two-fold, like force and guile in human affairs, and it may suit our plan or deceive us. For if I stay at home or do not stay, it can turn out very badly either way; but whatever happens, we are subject to the authority of the Prince. So it is also in games. Therefore, there are two kinds of happenings and non-happenings, one of them absolute and the other relative to plan or judgement.

Cardano was a compulsive gambler. He didn't believe that gambling was a zero-sum game – 'the path into error is always steeper and the loss is greater than the gain'. As any gambler knows, the low of losing is a larger feeling than the high of winning. Death, after all, is certain. The messiah's arrival is not, and nor is winning the lottery, and neither is perpetual life.

'Now, in general, gambling is nothing but fraud and number and luck,' Cardano wrote in *The Book of My Life*. Fraud one could come to recognise, and therefore avoid. Luck is out of one's control, although the gambler can learn not to be emotionally affected by its vicissitudes, and therefore lose the composure that Dostoevsky longed to possess – 'It is because fortune is adverse that the die falls unfavourably, and because the die falls unfavourably, he loses, and because he loses he throws the die timidly.'

A mathematician could do something about the numbers.

Cardano calculates the odds of throwing rolls of the dice, and combinations of dice, to derive frequency tables. He calculates the odds in drawing cards for the game of primero, which was a prototype of poker played with a deck of forty cards (stripped of 8s, 9s and 10s) and is, Cardano says, 'the noblest of all'.

In a game of dice or primero, or poker, the event is discrete, independent. A hand of poker begins with the dealing of cards, it ends when the deck is being shuffled in preparation for the next deal. It has a discernible beginning and end. Information is finite, the probabilities can be calculated. For statisticians and probabilists, a philosophical position has to be established, whether all life events can be analysed in this way, or if there will always be an area of unknowing. This divides us into classicists or subjectivists, believers in 'Laplace's Demon' or Charles Sanders Peirce's 'tychistic hypothesis'.

The study of statistics originated in an attempt to aggregate individuals into more predictable data, with a faith that the facts of the future are knowable from the experience of the past. It's the old argument: 'chance' only exists because of incomplete information. If we know everything, if we have sufficient data, enough numbers, then everything will be predictable.

Blaise Pascal attempted a 'geometry of chance' (*Aleae Geometria*), a set of procedures to map the processes of the world, and to account for them. A description of the 'great machine' of man and Earth and mathematics would enable a mastery of blind Fortune. 'Using geometry, we have so surely reduced it to an exact art that it shares its certainty and is ready to move boldly forward.'

Condorcet, or, to give him his full name and rank, Marie Jean Antoine Nicolas de Caritat, Marquis de Condorcet, presented a system of 'Social Mathematics' to soothe divisions in post-Revolutionary society – 'reason would triumph over words, truth over passions, enlightenment over ignorance'. Condorcet's social mathematics, like Pascal's geometry of chance and Buffon's moral arithmetic and Adolphe Quetelet's later social physics, presumes a 'common measure' – standardised weights and measures, universal public education, a uniform code of conduct, deduced from 'the general principles of natural law'. There is a bedrock of truth upon which we rely: all our judgements, including those upon which we act, are based on an intuitive reckoning of probabilities by 'vague, almost mechanical feelings', and this picture of the human species – freed from all its shackles, no longer dominated by chance or by the enemies of its advances, and striding with a firm and sure step along the path of truth, virtue and happiness...

You can almost hear the gulp in his voice, writing while nervously looking over his shoulder at the advancing mob. Condorcet was guillotined in 1794. *L'Esquisse d'un tableau historique des progrès de l'esprit humaine*, in which these hopeful words are written, was published posthumously the following year.

The determinist world view, which insists that all things are knowable, and predictable, and perfectible, continued with Laplace's Demon and Adolphe Quetelet's 'Social Physics'.

In his *Philosophical Essay on Probabilities*, Pierre-Simon Laplace wrote,

Given for one instant an intelligence which could comprehend all the forces by which nature is animated and the respective situation of the beings who compose it – an intelligence sufficiently vast to submit these data to analysis – it would embrace in the same formula the movements of the greatest bodies of the universe and those of the lightest atom; for it, nothing would be uncertain and the future, as the past, would be present to its eyes.

This is what has become known as Laplace's Demon. It's all about vantage point and data. The theory imagines an observer who has complete information about the present moment of any system, which therefore enables them to make accurate predictions of its future as well as full descriptions of its past.

Quetelet's Social Physics is based around his notion of *l'homme moyen*, 'the average man'. The average man seeks 'to avoid subjecting himself to the twitch of fortune's wheel' by positioning himself near the axis, 'thereby moderating the fluctuations, and preserving his tranquillity'. Averageness means an avoidance of extremes, and an avoidance of difference. The mean is perfect. Extremity is suspect. Variation is flawed, the product of error, morally and physically wrong. Everything should aspire to being the same. 'Deviations more or less great from the mean constitute ugliness in body, vice in morals, and a state of sickness with regard to the constitution.'

At the root of this disgust is the Spinozan notion that 'Nothing in nature is contingent but all things are from the necessity of the divine nature determined to exist and to act in a definite way'.

Without luck, where does variety come from? Where can it come from? We need the seeming benediction of chance, of luck... Epicurus's swerve, my father's unseen saviour in that Warsaw square. Darwin's theory of natural selection proposes that variety in nature is the consequence of chance encounters, changes and collisions. A gene mutates. One sperm in a hundred million reaches the egg (tonight I got lucky.) As Condorcet admitted, writing of Isaac Newton, 'a great discovery can arise from a fortunate combination of chance events and the efforts of genius'.

*

The English logician John Venn, in his 1866 book *The Logic of Chance*, disapprovingly quotes the late Professor Donkin, who says 'It will, I suppose, be generally admitted, and has often been more or less explicitly stated, that the subject matter of calculation in the mathematical theory of Probability is quantity of belief.'

Venn was a churchman, who saw no place for chance in the order of things, and no room for subjectivist thinking where God or mathematics might be concerned, 'The science of Probability... is simply a body of rules for drawing inferences about classes of events which are distinguished by a certain quality.'

Which begs the questions of who is drawing the inferences and how they might be applied, as well as how that 'certain quality' might be recognised.

Nietzsche reminds us that everything is in flux. 'Mankind is not a whole: it is an inextricable multiplicity of ascending and descending life-processes – it does not have a youth followed by maturity and finally by old age; the strata are twisted and entwined together.'

Everything is change; as Heraclitus says, we never step into the same river twice. All these currents and threads twist and flow into each other and away. Nothing is fixed, especially us. Your skin peels, your hair sheds, the starlight that touches your eyes at night was emitted from a body that died centuries ago.

The early twentieth-century pragmaticist philosopher Charles Sanders Peirce, who was an influence on Frank Ramsey, wrote that Chance is 'that diversity in the universe which laws leave room for, instead of a violation of law, or lawlessness'.

Chance is not just a kind of synonym for diversity. Peirce's universe is the same as Epicurus's (see Chapter 11), a swerving, messy place, where we clatter around with 'the unruly, the exceptional, and the disorderly'. It's not a matter of insufficient data, that we can't perceive an event's cause because we don't know enough or we're not situated at the correct vantage point. Peirce disputes 'whether it is exactly true... [that] every event has a cause... may it not be that chance, in the Aristotelean sense, mere absence of cause, has to be admitted as having some place in the universe.' His tychistic hypothesis is a version of the atomists' swerve, suggesting that 'the universe is constantly receiving excessively, minute, accessions of variety'.

The subjectivist tradition that stems from Peirce denies Laplace's Demon and Quetelet's Average Man, and all their certainties.

There is no external measure for calculating 'real' values, of hope or fear, or even risk. There is merely an observer, who devises and applies an algorithm, and, as the late Professor Donkin says, what the algorithm is measuring is the observer's own belief.

Doris Lessing in the second volume of her autobiography tells the story of being taken out for lunch by a publisher, who is trying to explain why it isn't worth putting any money into advertising her latest book. As an analogy she cites an exercise put to American military cadets. Put yourself in the role of a general. You have troops fighting on three separate fronts. On Front A, they're winning; Front B, holding their own; Front C, they're losing. You have one set of reinforcements. To which front do you send them?*

Put it another way: three patients are diagnosed. A seems to be recovering; B is holding their own; C is dying. Which is treated first?

These are the medical choices that govern our treatments. Practitioners calculate hazard ratios; they conduct risk assess-

---

* The correct answer is A. This might not seem humane, leaving one battalion to die, another with its destiny in the balance, but if you have any troops that are winning, the strategy taught to cadets is to enable them to win faster. The same with publishers. The most money spent on promotion goes on the books that are guaranteed to be bestsellers anyway. Win faster. 'All publishers think this way,' Lessing says.

ments and probable outcomes for anything from ulcers, on Q-scales, to heart attacks and strokes, on QRISK®.

The subjectivist statistician David Spiegelhalter points out that the average age of someone entering a UK hospital intensive care ward in 2020 with Covid-19 was sixty (the troops on Front A), whereas the average age of a Covid-19 death was eighty-two (Front C).

There has always been triage, he says – a sifting between admissions cases to determine who will benefit most from the treatment while not being harmed by it.

We talked by Zoom, because this was deep in the third Covid lockdown, and Spiegelhalter was taking time out from his unofficial role of explaining statistics to the nation on television and radio, as well as from his official role as Winton Professor of the Public Understanding of Risk at the University of Cambridge.

'You cannot separate probability from action and from consequences,' said Spiegelhalter, a passionate Ramseyite with a vocation for explanations,

> And it's crucial to understand that it's stratified. You cannot tell a person their risk. All you can say is that for a hundred people like you this is what we would expect. You're being embedded in a group that is sort of like you and sort of not like you.

So what does probability actually measure?

> The only thing it measures is us. The event itself does not have a probability, except possibly at the subatomic level. It's not the thing itself. It's not the 'truth'. We construct a measurement. It's the classic 'all models are wrong but some are useful'.

You're not measuring the concept, the concept doesn't actually exist – the map is not the territory, which is unfathomable and unmeasurable and unknowable. You're choosing what to measure and calling it something. It is not an objective aspect of the world. It's a way to operationalise a belief.

*

I have tried to calculate my father's odds of surviving the years 1939 to 1944.

A male Jew in Warsaw in 1939 had maybe a 10 per cent chance of surviving the German occupation.

A foreign slave labourer in the Gulag in 1940 maybe 50 per cent.

A Polish soldier at the Battle of Monte Cassino in 1944, 60 per cent?

In this case, the observer decides to ignore the dysentery and malnutrition his father was suffering from at other times during those years, and the other battles and ordeals and misfortunes he endured along the way. To convert this into a probability, we translate the percentages into decimals and multiply them: $.1 \times .5 \times .6 = .03$ or 3 per cent. So, Joe Flusfeder had a 3 per cent chance of surviving the events in Europe as they touched him in those years of war. To have a 97 per cent chance of not surviving are not acceptable odds. The practitioner would discontinue treatment in favour of finding another patient with a better chance of survival.

Except, that isn't quite right. That would be the way to calculate events that are independent of each other. Hands of cards are independent of each other. Few life events truly are. My father had to get out of Warsaw to be allowed a chance of surviving Siberia. He had to have survived Warsaw and Siberia to be allowed his chance of getting through the Battle of Monte Cassino. A more accurate estimation of his odds would use a different formula, calculating each of its parts in the event of the preceding part, which shrinks the chances to beyond infinitesimally small.*

In 2001, my father went into hospital for what was described as 'a routine procedure'. His carotid artery was occluded, and it was necessary to clear it. There was, the surgeon told us, a 95 per cent chance of everything going perfectly.

It did not go perfectly. After the general anaesthetic wore off, we – my stepmother, stepbrother and I – were called into the recovery ward.

'Is English not his first language?' they said. 'No, Polish,' we said. 'We think he might be speaking Polish. Maybe it's an effect of the anaesthetic,' they said.

He was not speaking Polish. He was talking in nonsense syllables, sometimes with arm gestures for greater emphasis. In the course of the procedure (at no point was it called an 'operation'), some 'matter', fragments of the plaque that was blocking his artery, had migrated to his brain and ended blood supply to an area responsible for producing and comprehending language. It was evident on the X-ray they took the following day – a part of the Wernicke's area was greyed out where the cells had died.

For maybe the first time in his life, at the age of seventy-nine, Joe Flusfeder got unlucky.

---

* Instead of the probability (p) being derived by $p(A)\, p(B)\, p(C)$, we would have to use the formula $p(A \mid B, C)$, $p(B \mid C)$, $p(C)$, in which each term is calculated in relation to the others – the vertical line means 'given that'.

*

Frank Ramsey, Wittgenstein's translator and intellectual comrade and occasional adversary, who unluckily died of a liver infection at the age of twenty-six but has subsequently influenced much thinking in logic, probability and economics, was a follower of Donkin rather than Venn. Ramsey developed the subjectivist line in probability. Beliefs, he said, are bets.

> Whenever we go to the station we are betting that a train will really run, and if we had not a sufficient degree of belief in this we should decline the bet and stay at home.

All probabilist thinking does is enable you to make estimates. 'It's the logic of partial belief and inconclusive argument,' Ramsey wrote. And even if it is impossible to identify belief with an actual value, money can still be a way to keep score. Ramsey was a poker player. In a letter to Lytton Strachey about a Bloomsbury poker game, Ralph Partridge wrote, 'Frank, with the guffaws of a hippopotamus and terrible mathematical calculations, got all our money from us'.

Frank would have got my money from me. It felt as if anyone could get my money from me. I was the mark, the sucker, the tourist, the provider, the fish. Come play with David, he'll pay off your good hands, and even if he sees that you're bluffing and calls you all the way down, your miracle out card will hit on the river. At least that's how it felt. At least that's what I believed.

*

Buffon's first mathematical work, winning him admission to the French Academy of Sciences in 1734, was an analysis of *franc-carreau*, which is a version of the roll-a-coin game that Wittgenstein played at Midsummer Common two hundred years later. Using calculus, Buffon analysed the odds of a coin landing fully inside a square and not across any of the dividing lines.

Wittgenstein made his choice not to influence the passage of the penny on its journey to land inside a prize-winning square or on its lines. Buffon analysed the odds, and they were there for any subsequent player to see. Wittgenstein would have known this. The later philosopher was an enthusiast for Buffon's *Discours à l'Académie Française* (also known as *Discours sur le style*), its advice on good writing, and its relationship to character.

'Style is the man himself,' Buffon wrote, an idea echoed by Wittgenstein in his notebooks,

> Genius is talent in which character makes itself heard. ... This is no mere intellectual skeleton, but a complete human being. That too is why the greatness of what a man writes depends on everything else he writes and does.

There are questions of authenticity here (Wittgenstein would probably have called it 'decency'). For Buffon, style is something immutable, particular to the individual writer, embedded. Voice and character and subject are all of a piece,

> Only those works that are well-written will pass down to posterity. The quantity of knowledge, the singularity of the facts, even the novelty of discoveries, will not be sure guarantees of immortality; if the works that contain them are

concerned with petty objects, or if they are written without taste or nobility... Style is the man himself [*Le style c'est l'homme même*]; style cannot be stolen, transported, or altered; if it is elevated, noble, and sublime, the author will be admired equally in all times, for only truth is durable and everlasting.

And 'truth' here might be what passes for selfhood.

If you can weigh everything, if you can ascribe a moral value to every event, if you can measure, as Buffon said, 'the real value of our fears and of our hopes', then you can organise your material – whether it be a taxonomy of geological strata or a book about luck – according to the principle you choose, weight or height or significance.

But for those of us who believe, with Spiegelhalter and Ramsey and Peirce and the late Professor Donkin, that these are notional expressions of our beliefs rather than an absolute measurement of an external thing or event in the world, then you are thrown back against questions of character and self, the writer's. And if you don't truly believe in a self that's fixed? Maybe my struggles – with this book, at the poker table – are all to do with that being unresolved.

Nietzsche told us to *Become what we are, once we know what that is...* What if we don't know what that is? Maybe the 'what we are' isn't fixed either? How then do we make our choices over how to present the material?

I could follow Cardano's principle in his memoir, to lay everything out chronologically – 'I have simply put them down in the order in which they happened,' he writes. But even that procedure is suspect, because who is to say when and how an event begins? As Cardano concedes, 'the beginning and outcome of important events are not always evident'. Most events are dependent.

My book is not a memoir, therefore the chronological principle does not apply. When I set out on this project seven years ago, I might have listened harder to Buffon,

> It is for lack of plan, for not having sufficiently reflected on his purpose, that even a man of thought finds himself confused, and knows not where to begin to write; he perceives at the same time a great number of ideas; and since he has neither compared nor arranged them in order, nothing determines him to prefer some to others; he remains perplexed. But when he has made a plan, when once he has assembled and placed in order all the essential thoughts on his subject, he will perceive at once and with ease at what point he should take up his pen; he will feel his ideas ripening in his mind; he will hurry to bring them to light, he will find pleasure in writing, his ideas will follow one another readily, his style will be natural and easy; a certain warmth will arise from this pleasure, will spread over his work, and give life to his expression; animation will mount, the tone will be elevated, objects will take colour, and feeling, joined to light, will increase and spread, will pass from that which we say to that which we are about to say; the style will become interesting and luminous...

I know what the next chapter is going to be, we're moving towards it now, so there is a kind of teleology at work, this chapter works its way towards the next, as the previous one moved towards this, in the order determined by the randomiser. If the algorithm had been different; if, say, I had 'chosen' each succeeding chapter only when I had reached the end of the current one, its ideas 'ripening' in my mind, Chapter 9 probably wouldn't have found its late emphasis on style and structure – and its almost silent subtext.

This book is in part a movement from darkness into light, from superstition to rationality, bringing its author, me, into a closer apprehension of his own character.

I have never been sure of my plan. I hope by the end I will have assembled 'all the essential thoughts' on my subject. Maybe then I will know how to begin.

A couple of years ago, I went to Weimar, to visit Goethe's house there, and the statue of Agathe Tyche he designed for his garden. It is a sphere balanced on a cube, which represents the internal dynamics of fortune and might be the first conceptual, non-representational sculpture. Schiller lived in Weimar too, a walk away from Goethe's house, and I was going to write about both of them, and Goethe's unbuilt second Tyche-inspired statue. Most of this material would have been in the previous chapter, because Goethe visited Vesuvius, and wrote a travel journal of his experiences there.

I've cut out the Weimar material, in the interest of flow and narrative and style. One of my problems in writing this book has been to decide what to exclude, because almost everything I have ever written belongs to the unavoidable subject of luck.

In the late 1990s, I interviewed the scholar Rafael ('Felek') Scharf. Scharf was a Polish Jew, from Cracow, who had left Poland in 1938. He had been in British military intelligence during the war, had careers as a silkscreen manufacturer, as a dealer in English watercolours, and after his retirement devoted himself to exploring, memorialising, and trying to heal the problematic phenomenon of Polish Jewishness, which remained, to his death, despite everything, his unit of identity. He published one book, *Poland, What Have I to Do with Thee?*, a collection of essays, feuilletons and reviews, which occupied itself on this subject with great tenderness and rigour. Every word had been considered, each sentence had the force of moral truth.

In the penultimate essay in his book, Scharf celebrates his Cracow Hebrew schoolteacher Benzion Rappaport. Rappaport was an inspirational teacher, of ethics as much as religion. What was most important to him, even if it risked him his job, was to inculcate 'the spirit of free-ranging, open-minded enquiry'.

> He took me aside and what he told me I have never forgotten.
>
> 'Dear boy,' he said… 'The most important thing is the question man has to put to himself when he raises his eyes to heaven. *Ma chovato b'olamo* – what is my duty in this world?' Every morning, before you begin your day, ask yourself this question – but seriously, not just casually. Every day afresh – and think about it a minute. Do not try to answer it – there is no short answer to it, it will not come to you quickly, maybe it will never come to you – it matters not. The thing is to realise that the question is important, that you have a duty to perform and have to search for it.'

After the war was over, news came to Scharf, by then a sergeant in the British Intelligence Corps, that his mother had survived the war, the only other member of his family to do so. He caught a plane to Germany, where he requisitioned a car and drove towards Cracow, stopping off at Warsaw. One of the few undamaged buildings was the Hotel Polonia, where the British Embassy had its headquarters. Scharf pushed through the usual crowd, of soldiers and diplomats and dealers and petitioners, and – 'a heartstopping moment this' – he spots a familiar face, an old schoolfriend from

Cracow looking for survivors. They embrace, they talk feverishly, they exchange information. An elderly Polish peasant who has been observing them for a while approaches. He asks if they are Jews. They admit, perhaps uneasily, that they are. The old man takes out from his coat a bundle of pages from an exercise book, covered in faded handwritten Hebrew script. With it is a topsheet scrap of paper, a scrawled message in Polish, *Pobożna dusza. To jest dzieło życia człowieka. Oddaj to w dobre ręce.* 'Pious soul. This is a man's life work. Give it into good hands.'

The two reunited friends look at the pages. They immediately recognise it as the work of their old teacher Rappaport, a collection of essays on German philosophy, Hegel, Kant, Schopenhauer, as well as Rappaport's own thoughts on religion, ethics, the method of scientific enquiry. He'd thrown the book out of the window of the train taking him to the death camp in Belzec. And, in an arbitrary open field, a man, the one who now stands in front of us, finds it, deciphers the Polish message, safeguards the manuscript. When the war ends, he travels to Warsaw to look for Jews to hand it over to. They are hard to find, but in the crowded lobby of the Hotel Polonia he spots two Jews – two former pupils of Benzion Rappaport.

A marvellous coincidence, from which the agnostic derives his faith.

poker –£1,243

CHAPTER 10

By Paths Coincident (canned chance)

The greatest, grandest things are unpredicted.

Herman Melville, marginal note on copy of
Milton's *Paradise Lost*

If you look upon an old wall covered with dirt, or the
odd appearance of some streaked stones, you may
discover several things like landscapes, battles, clouds,
uncommon attitudes, humorous faces, draperies, etc.
Out of this confused mass of objects, the mind will be
furnished with an abundance of designs and subjects
perfectly new.

Leonardo da Vinci, *A Treatise on Painting*

net total –£1,286

This chapter is going to look at how art and literature have
incorporated ideas of luck, in both theme and structure.

George Eliot's novel *Daniel Deronda* opens in a German
casino in August 1865. Dostoevsky is probably there, among
the 'very distant varieties of European type' surrounding the
roulette tables, where Gwendolen Harleth is on what poker
players call a heater. She is winning astonishingly, coup after
coup. 'She had begun to believe in her luck, others had begun

208

to believe in it; she had visions of being followed by a *cortège* who would worship her as a goddess of luck…'

It might be the attention of Daniel Deronda upon her, or the reader's, or it being the nature of luck to shift and change, but the wheel turns, all the money in front of her rushes away; she is no longer the goddess of luck. 'Since she was not winning strikingly, the next best thing was to lose strikingly.'

Nothing seems really to be at stake. Gwendolen is on tour, everyone is admiring her, her family is monied. Except, as she finds out that evening, she is suddenly poor. The company administering the family fortune had 'also thought of reigning in the realm of luck'.

The money that underpins her life is lost – bad speculations, a trusted family adviser who turns out to have been a crook. All that 'overseas' money, in this case Gwendolen's grandfather's Barbados estate, and which generally in eighteenth- and nineteenth-century English fiction means sugar and slaves, is gone. The only way out for her is a 'good' marriage.

Here is the Victorian age, and ours. The foundations are chanceful, and usually corrupt. Fortunes are built on speculation and other people's suffering. And it can all go at any moment. One of the stock characters of Victorian novels is the charismatic swindling speculator (Dickens's Mr Merdle, Trollope's Augustus Melmotte) who rises like a comet to the heights of society before burning out, the fire consuming both himself and all those who trusted in him. The casino itself was a matter for polite disapproval (in Eliot's earlier *Felix Holt*, the eponymous main character warned that the working classes were 'taking up the worst vices of the worst rich'), but everything is founded on the uneasy ground of speculation. The year that *Daniel Deronda* was published, 1876, was one of the few when George Eliot made more money from her fiction than from the stock market; four years later, she married her financial adviser.

Eliot (the pen name of Mary Ann Evans) understood the fear of losing it all, and the psychology of the debtor, the one who has so little and owes so much that the world seems entirely against them, and they long for change,

> The temptation, particularly when young or otherwise relatively powerless, is to exchange one dependency for another. Better the sudden random act than a parent's whim or negligent selfishness... for the majority, who are not lofty, there is no escape from sordidness but by being free from money-craving, with all its base hopes and temptations, its watching for death, its hinted requests, its horse-dealer's desire to make bad work pass for good, its seeking for function which ought to be another's, its compulsion often to long for Luck in the shape of a wide calamity.

This is from *Middlemarch*, Eliot writing as a student of Kant, following his notion of *Unmündigkeit*, the unagented timidity which her more frustrated, self-defeating characters find themselves constrained by.

Gwendolen in *Daniel Deronda* would aim higher than her pinched post-casino world, where a bad marriage to a cruel landowner is her only way out.

> Other people allowed themselves to be made slaves of, and to have their lives blown hither and thither like empty ships in which no will was present; it was not to be so with her.

She doesn't like to hear her mother's submissive remark, 'We must resign ourselves to the will of Providence, my child...'

Mrs Davilow might have been speaking for all characters in eighteenth- and nineteenth-century fiction, where chance meetings, coincidental events, abrupt shifts of fortune are presented as the action of providence, as authorial plan and divine intention become one (see, for example, *Tom Jones* or *Mansfield Park*). The Tychists among us might read a contradiction here, between the supposedly chanceful foundations of the characters' worlds and the narrative inevitability of their fates.

This continues into the twentieth century: the beginnings of *Robinson Crusoe* and *The Great Gatsby* are in effect the same. 'It was a matter of chance that I should have rented a house in one of the strangest communities in North America,' says *Gatsby's* narrator Nick Carraway. 'But being one Day at Hull, where I went casually, and without any Purpose of making an Elopement that time; but I say, being there, and one of my Companions being going by Sea to London, in his Father's Ship, and prompting me to go with them…' writes Crusoe.

Nick Carraway rents a house in West Egg so he can be neighbour to Jay Gatsby and bear witness to his glamour and tragedy and narrate the events of *The Great Gatsby*. Crusoe goes to sea so that he might be shipwrecked and write a journal of his survival on a desert island. Crusoe's *casually* and Carraway's *chance* have the same root, *cadere*, meaning accidental, or fortuitous. All is as it was meant to be, how things fall, the providence dictated by author and God to characters as resigned as Mrs Davilow or as full of seemingly self-determining rage as Captain Ahab in Herman Melville's *Moby Dick*.

Moby Dick's narrator is Ishmael, who falls in the tradition of following what was meant to be, 'And, doubtless, my going on this whaling voyage, formed part of the grand programme of Providence that was drawn up a long time ago.'

His captain, Ahab, seems to be self-consciously in this tradition too. He cries to the gods,

Come, Ahab's compliments to ye; come and see if ye can swerve me. Swerve me? ye cannot swerve me, else ye swerve yourselves! man has ye there. Swerve me? The path to my fixed purpose is laid with iron rails, whereon my soul is grooved to run. Over unsounded gorges, through the rifled hearts of mountains, under torrents' beds, unerringly I rush! Naught's an obstacle, naught's an angle to the iron way!

Here is the machine voice of the industrial revolution, the captain at sea as dauntless as any steam locomotive making its necessary route from origin to destination through whichever part of nature is unlucky enough to stand in its way.

But Ahab in his urge for cataclysm turns his back on technology, on Newtonian physics, on modernity; he tramples his quadrant and curses science, impatient with induction's inability to guarantee 'where one drop of water or one grain of sand will be to-morrow'. Instead, to the mast of the *Pequod* he nails a doubloon, that figure of chance, the coin-flip.

The captain is the embodiment of the limits of induction, an exemplar of Ramsey, of the early Wittgenstein. He tries to resist teleology – that every event is moving towards its own consummation – rejecting Christian notions of providence, of submission before manifest destiny, the processes that Ishmael is sufficiently humble to recognise, 'chance, free will, and necessity – nowise incompatible, all interweavingly working together'.

The central necessity here is in the reader's knowledge of the writer's intention that Ahab, the mad hunter, and his quarry, the great white whale, must meet. The same

implacable principle is enacted throughout eighteenth-
and nineteenth- and early twentieth-century literature and
beyond. Providence and necessity and narrative form and
authorial intention all combine – if they don't start off as
aspects of the same thing, they become it. This occurs most
nakedly in Thomas Hardy's poem 'The Convergence of the
Twain'. Here, the routes of the *Titanic* and the inevitable
iceberg are described in alternating three-line verses until the
two protagonists meet, as they must do, as the form of the
literary work demands that they do,

> By paths coincident
> On being anon twin halves of one august event

Writers have a weakness for symmetry. The end of *Daniel
Deronda* echoes its beginning. Now it is Deronda who
is prepared to risk everything, in an existential casino of
history and identity. He has thrown away his class position
by 'coming out' as a Jew. Turning his back on the propertied
state of an English gentleman, he will make a trip to the East,
to implement a messianic Zionism, gambling that by striving
to give a dispersed nation a country, he might heal the world.

But what if we're Gwendolenians or Ahabians at heart? If
we're no longer content sharing the submissiveness of Mrs
Davilow under the heavy symmetries of great novelists? If
we grow impatient with representations of divine will and
narrative necessity? What if the reader wants out?

*

I can still remember the thrill of my first encounter with Julio
Cortázar's *Hopscotch*. I came across the book by chance in
a bookshop in Fort Lee, New Jersey, where I would go to

browse while visiting my father and stepmother on difficult teenage summer holidays.

The Book Cave was run by two women, one of whom was maybe in her twenties and the other in her thirties, but I can hardly be sure of that because, in my time as a customer at the Book Cave, I was between the ages of twelve and fifteen and my perceptions of many things were unreliable.

I was shy, and very thin and very pale, and the only things I could be sure of were books, and the worlds they contained. Most days I would visit the Book Cave, to choose what of world literature I was next going to consume. Possession, when you can count up the things you own, is a magical act. Each new book I bought was an act of identification so strong that it was a kind of incorporation.

Some of the books I bought then are still with me: Ralph Ellison's *Invisible Man*, Thomas Pynchon's *Gravity's Rainbow* – which elicited some concern from the younger bookseller, who wondered if I was quite ready for that one; and she was right: I wasn't. And *Hopscotch* by Julio Cortázar.

Hopscotch (*Rayuela* in its original Spanish) was published in 1963, and translated into English in 1966. It is probably a young person's book. Argentinian émigrés in 1950s Paris have long arguments about art and philosophy. It rains. They fall in and out of love to a jazz soundtrack. The book itself is in love with the modern city and chance collisions. The narrator returns home and disintegrates over the course of the, increasingly fragmented, novel whose form provides a more reliable cohesion than the narrator's consciousness.

Prefacing the book is a 'table of instructions' in which the author informs us that 'this book consists of many books, but two books above all. The first can be read in a normal fashion and it ends with Chapter 56… The second should be read by beginning with Chapter 73 and then following the sequence indicated at the end of each chapter. In case of confusion or forgetfulness, one need only consult the following list.'

> TABLE OF INSTRUCTIONS
>
> In its own way, this book consists of many books, but two books above all.
>
> The first can be read in a normal fashion and it ends with Chapter 56, at the close of which there are three garish little stars which stand for the words *The End*. Consequently, the reader may ignore what follows with a clean conscience.
>
> The second should be read by beginning with Chapter 73 and then following the sequence indicated at the end of each chapter. In case of confusion or forgetfulness, one need only consult the following list:
>
> 73 - 1 - 2 - 116 - 3 - 84 - 4 - 71 - 5 - 81 - 74 - 6 - 7 - 8 - 93 - 68 - 9 - 104 - 10 - 65 - 11 - 136 - 12 - 106 - 13 - 115 - 14 - 114 - 117 - 15 - 120 - 16 - 137 - 17 - 97 - 18 - 153 - 19 - 90 - 20 - 126 - 21 - 79 - 22 - 62 - 23 - 124 - 128 - 24 - 134 - 25 - 141 - 60 - 26 - 109 - 27 - 28 - 130 - 151 - 152 - 143 - 100 - 76 - 101 - 144 - 92 - 103 - 108 - 64 - 155 - 123 - 145 - 122 - 112 - 154 - 85 - 150 - 95 - 146 - 29 - 107 - 113 - 30 - 57 - 70 - 147 - 31 - 32 - 132 - 61 - 33 - 67 - 83 - 142 - 34 - 87 - 105 - 96 - 94 - 91 - 82 - 99 - 35 - 121 - 36 - 37 - 98 - 38 - 39 - 86 - 78 - 40 - 59 - 41 - 148 - 42 - 75 - 43 - 125 - 44 - 102 - 45 - 80 - 46 - 47 - 110 - 48 - 111 - 49 - 118 - 50 - 119 - 51 - 69 - 52 - 89 - 53 - 66 - 149 - 54 - 129 - 139 - 133 - 140 - 138 - 127 - 56 - 135 - 63 - 88 - 72 - 77 - 131 - 58 - 131 -
>
> Each chapter has its number at the top of every right-hand page to facilitate the search.

There's an exhilaration of structure, a deadpan formal playfulness that still thrills. And while the young-man yearning ('Would I find La Maga?' is the opening line of 'the first' book – declaring the themes of randomness and loss and the secret patterns of city chance) doesn't have the same significance for me now as when I first picked it up in the Book Cave, I still love *Hopscotch*. It's the book that taught me most about reading. And, not entirely coincidentally, it's the book that made me realise I was going to become a writer.

At least two revolutionary implications follow from Cortázar's table of instructions. The first implication is that the reader may choose their own sequence for a 'third book', or the 'second book', or even for the first, or allow any random procedure to decide it. The second implication is that all books can be read in this way, whether they are designed to be or not, liberated from the determinings of plot and chronology and consequence and development, of character or mood or rhythm, of their authors' intentions and judgements and tastes. Things are contingent. They don't have to fall this way, even in something so seemingly fixed as the printed matter of a book. Everyone may choose their own adventure. Anything is possible. Everything changes.

*

Samuel Beckett served a kind of literary apprenticeship to James Joyce, acting as amanuensis to the older man as Joyce's sight continued to weaken. According to an anecdote delivered by Beckett to Richard Ellmann, when James Joyce was dictating *Finnegans Wake*, there was a knock on the door which Beckett didn't hear. 'Come in,' Joyce said. Later, when Beckett was reading back what had been written that day, Joyce interrupted, 'What's that "Come in"?' 'Yes, you said that,' Beckett said. Joyce took a moment to consider, and then said, 'Let it stand.' As Ellmann wrote, 'He was quite willing to accept coincidence as his collaborator.'

The closest I can come to finding that 'Come in' is on page 512 of *Finnegans Wake*, 'The house was Toot and Come-Inn by the bridge called Tiltass...' Beckett's anecdote suggests that the sentence began, simply, as an unornamented, 'The house was by the bridge...' and that the 'Toot and' was added subsequent to the chance phrase entering the book, to make the Tutankhamen pun.

Does allowing the possibility of chance operations into *Finnegans Wake* raise the work or lower it? 'Let it Stand' is the English translation of the Latin word 'stet', the mark traditionally made on the page by the author when rejecting an editorial emendation or suggestion. Which makes paramount the intentionality of the author, the Great Man theory of literature. 'A book,' Joyce said, 'should not be planned out beforehand, but as one writes it will form itself, subject... to the constant emotional promptings of one's personality.'

Beckett's own artistic principles were different, prompted by his sense of the formlessness outside rather than a concern with the management of material from within,

> What I am saying does not mean that there will henceforth be no form in art. It only means that there will be new form and this form will be of such type that it admits the chaos and does not try to say that the chaos is really something else. The form and the chaos remain separate. The latter is not reduced to the former. That is why the form itself becomes a preoccupation, because it exists as a problem separate from the material it accommodates. To find a form that accommodates the mess, that is the task of the artist now.

To find a form that accommodates the mess... It is not enough to acknowledge chance and luck as subjects, to adopt them as theme, with the longed-for cataclysm being a coup of plot. The 'unpredicted' elements in the story of *Moby Dick* don't convince as being exempt from plan, from grace, because the shape and structure of the novel, albeit in a very baggy form, follow the teleological rules as relentlessly as a Hardy poem.

Two novels published in 1969 exemplify the opposing routes one can follow to answer Beckett's question: you might choose a set of constraints, an algorithm, that will reach past the limited position of the artist to the formless messy greater truth; or go as constraintless as possible and hope that the absence of external limits will allow a fundamental sincerity to liberate the imaginations of writer and reader.

La Disparition by Georges Perec comes out of the procedural experiments developed by the largely French group of the Ouvroir de Littérature Potentielle (Workshop for Potential Literature), which was first established in 1960 by the writer Raymond Queneau and the mathematician François Le Lionnais. The members of OuLiPo enjoyed puzzles, games, acrostics, complications. *La Disparition* (literally 'The Disappearance', but whose English translation, a bravura performance by Gilbert Adair, is called *A Void*) is OuLiPo's long-form triumph, performed so deftly that an early reviewer failed to notice that the novel is a lipogram. *La Disparition*'s self-imposed constraint is not to contain any incidences of the letter 'e', which in French as in English is the most commonly used. The *e* in French is pronounced *eux*, and therefore sounds the same as 'them' or 'they'. Perec's missing letter refers in its absence to all those, including the author's parents, who were murdered during the German occupation of France during the Second World War.

Perec once wrote, 'Basically, I set myself rules in order to be free.' It is the 'aleatory' component in *La Disparition*, the element of game, with its rules and compunctions and necessities and exclusions, that enlarges the text and its emotional effect, the letter's absence allowing in an unsayable loss, and an unspoken grieving.

The English novelist B.S. Johnson published *The Unfortunates* that same year in 1969, a book in a box. It also is a novel about grief, but uses an opposing strategy. On the inside front cover is an explanatory note (the 1960s gave us, if nothing else, many explanatory notes) that says,

This novel has twenty-seven sections, temporarily held together by a removable wrapper.

Apart from the first and last sections (which are marked as such) the other twenty-five sections are intended to be read in random order.

If readers prefer not to accept the random order in which they receive the novel, then they may re-arrange the sections into any other random order before reading.

The section marked 'First' grounds us. A journalist is sent to a random town to report on a football match; he comes out of the train station and realises he's been here before – 'I know this city!' It's where Tony used to live, his close friend who died recently of cancer. 'Last' occurs after the game is over, the writer back at the station, his scattered thoughts on the way home, preoccupied with love and even more so by death, Tony's by illness, his own, imagined, by suicide, in front of the train he is waiting for. In the twenty-five unordered sections in between, reminiscences of Tony collide with snippets of travel to the match, and the drafting of the report, which the narrator finally phones in to the newspaper copytaker. The report, 'from BS Johnson', of City 1 United 0, as if cut out of a newspaper, is on the inside back cover of the box that contains *The Unfortunates*.

Johnson did kill himself, four years after publication. 'His real enemy,' the critic Frank Kermode later wrote, 'was not what he thought of as the inevitable falsity of stories but an agonised egotism, the sense that it was essential but impossible to tell the whole truth about himself.' Johnson's original plan was to have no fixed chapters but he lost his nerve or his courage in the face of publishing caution. A result is that the gimmicky nature of *The Unfortunates* is emphasised, and the 'experiment' or 'adventure' can read as an over-literal attempt to mimic the seemingly arbitrary movements of human consciousness, or at least its author's.

The copy of the book that I looked at in the British Library has, by the side of the explanatory note, a handwritten dedication from the author to his friends, the writer Zulfikar Ghose and the artist Helena de la Fontaine, 'For Zulfi and Helena, Oldest and best, with much love, this ~~book~~ novel which cost me so much pain.'

This dedication lifts the novel, granting an emotion that I'd missed on previous readings of *The Unfortunates*. It might be the signature that follows, 'Bryan', scrawled with a flourish, the author's first name released from its customary hidden place behind its initial letter; or maybe the frank admission of the pain it cost him to make this thing, which shames me for having reduced to a gimmick this work that must have been for its maker the most important thing in the world, and utterly intractable; or maybe it's the way Johnson had crossed out the word 'book' and replaced it with 'novel' because disarranged pages in a box aren't quite the same thing as a book, and the correction gives us access into the mind of its writer in a wavering, doubting instant. This is the feeling moment that Roland Barthes called the *punctum*, 'the sting, speck, cut, little hole – and also a cast of dice... that accident which pricks me (but also bruises me, is poignant to me).'

*

Long before the 1960s, visual artists were enacting their own swerves, embedding their investigations of chance into the nature and structure of the work of art.

Marcel Duchamp (who would later accept the position of OuLiPo's American correspondent) first systematised the use of chance procedures to make artistic works.

In 1913, Duchamp made his '3 stoppages étalon' ('3 standard stoppages') which, in its ordinariness, its deadpan silliness, invokes and mocks the Cain-tradition of Enlightenment measurers. Duchamp held a one-metre-long thread, 'straight and horizontal', one metre above a blank canvas, before letting it fall. With a layer of varnish he fixed the chance shape the thread made upon landing, and repeated the process two more times. Using the varnished shapes as templates, he constructed three rulers, with which he drew the three lines of the finished piece,

3 standard stops =
canned chance
…
a new image of the unit of length

The metre had been defined, or 'measured', as one ten-millionth part of the distance from the North Pole to the Equator.

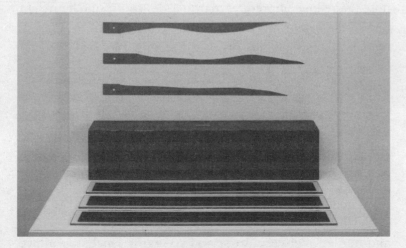

Photo of Marcel Duchamp '3 stoppages étalon 1913–14' © Tate/ Association Marcel Duchamp/ADAGP, Paris and DACS, London 2021

But because the Earth is not a perfect sphere, the so-called standard metre is shorter than it 'should' have been and, therefore, essentially as arbitrary as Duchamp's own yardsticks. He instructed that his stoppages were to be displayed in a box 'to preserve them from other measurements', as if the Enlightenment and its subsequent tradition of standardised results acted as a kind of contagion.

That same year, Duchamp carried his chance procedures into musical composition.

'Erratum Musical' was composed by writing twenty-five musical notes on twenty-five separate pieces of paper, which were then randomly selected, with each note written down in the order of its selection. This was again repeated two more times, to produce three melodies, 'three different scores', which were performed by three voices, originally those of Marcel and his sisters Yvonne and Magdelaine. The lyrics were found by opening a dictionary at random. The siblings sang the definition of *imprimer* ('printing'), one syllable per note (Faire une empreinte; marquer des traits; une figure sur une surface; imprimer un scau sur cire, 'To make an imprint; mark with lines; a figure on a surface; impress a seal in wax').

*

Sometimes art can change the world. Taking objects as chance offered them – a musical note, the shape of a thread, or his 'readymades': a bicycle wheel, a bottle dryer, a comb, a shovel, a urinal – and signing them and calling them art was intended not to exalt the artist, but to enable art to exist everywhere, for everyone, to break the boundary between art and non-art, between artist and spectator. Duchamp aimed to go beyond pretty 'retinal' sensations, the deft copying of things in the world to trigger a predictable aesthetic response in the viewer. 'For the spectator even more than the artist, art is a habit-forming drug.'

'Lucky or unlucky chance is a completely personal matter,' he told his chronicler Arturo Schwarz. 'My chance is not the same as yours: what is lucky for one person may be unlucky for somebody else. I was interested in expressing this concept physically.'

Duchamp got here before me, investigating the personal nature of luck, and its relationship to time. Kairos is present in his flashes of messianism, his 'ministry of coincidences', his work 'planning for a moment to come (on such a day, such a date such a minute)... The important thing then is just this matter of timing... It is a kind of rendezvous'.

And he gambled. Duchamp made the '3 stoppages' by collaborating with gravity. In other works he made collaborations with electricity and wind. In 1925, Duchamp went to Monte Carlo to play roulette. His principal collaborators now were money and time, as well as chance, as always.

Before leaving Paris he issued a series of bonds to finance the trip.

Extracts from the Company Statutes

Clause No. 1. The aims of the company are:

1. Exploitation of roulette in Monte Carlo under the following conditions.

2. Exploitation of trente-et-quarante and other mines on the Côte d'Azur, as may be decided by the Board of Directors.

Clause No. 2. The annual income is derived from a cumulative system which is experimentally based on one hundred thousand rolls of the ball; the system is the exclusive property of the Board of Directors.

The application of this system to simple chance is such that a dividend of 20 per cent is allowed.

Clause No. 3. The company shall be entitled, should the shareholders so declare, to buy back all or part of the shares issued, not later than one month after the date of the decision.

Clause No. 4. Payment of dividends shall take place on 1 March each year or on a twice yearly basis, in accordance with the wishes of the shareholders.

Duchamp set out to raise 15,000 francs from 30 bonds of 500 francs apiece. It is unclear how many bonds he actually sold, or the betting system he used in his two months in Monte Carlo. He called it a Martingale, a term that originally referred to a specific betting procedure of doubling one's stake at every loss, reverting to the original stake in the event of a win – an infallible system so long as the gambler possesses an infinite bankroll to withstand an extended losing streak. Duchamp however seemed to use the word to refer to any kind of betting system.

The Martingale is without importance. They are all either completely good or completely bad. But with the right number even a bad Martingale can work and I think I've found the right number. You see I haven't quit being a painter, now I'm sketching on chance... It's delicious monotony without the least emotion...

His friend, the artist Man Ray, helped him design the bond, with its photograph at the top of the artist's face covered in lather from soap and shampoo. The bond was co-signed by Marcel Duchamp and Rrose Sélavy (a homophone for *eros, c'est la vie*), his female alter ego. Duchamp liked assuming personas. He established his international reputation in 1917 by submitting a readymade urinal for a group show of modern artists in New York. 'Fountain' was signed by 'Richard Mutt', whom Duchamp conceived of as a woman artist posing as a man. Another of the names he went by around the time of his Monte Carlo adventure was the near-anagram Marchand du Sel, which means 'Merchant of Salt', or 'Merchant of Seal'. Duchamp was already making a profession of being an autographer – applying his signature, or 'seal', to any object to give it greater value. Quite soon, the bond became a collector's item, worth far more than the original investment.

In January 1925, before heading out to Monte Carlo, Duchamp wrote to one of the subscribers, Jacques Doucet,

I've just mailed the bond which we spoke about yesterday. It's self-explanatory. I have studied the system a great deal, basing myself on my bad experience of last year. Don't be too sceptical since this time I believe I have eliminated the word chance. I would like to force roulette to become a game of chess.

By June, the word chance had yet to be eliminated,

the game of chance becomes an exercise in discipline, a test of indifference and patience over the long period of random moves that follow one another without memory, and therefore without history. But for those who play without really playing, or who play at playing, without passion or emotion, without illusions of any kind, but with the concern of maintaining a form of economy of oneself (and of one's stake), the game of chance turns out the most boring of games.

Later that year, Doucet received a fifty-franc dividend, as promised.

*

With his readymades, and his 'multiples', and his preferences for gravity and wind and chance, and money and time, over skill or retinal appeal as collaborative partners, Duchamp was the artist of the twentieth century. Duchamp's examples of pictures of 'lucky or unlucky chance (in or out of luck)', which refused to be determined by habit or taste, inspired generations of the influenced and the imitating. Hans Arp composed chance scraps of paper into collages, shuffling and gluing them as they fell; Tristan Tzara's Dadaist poems followed Duchampian principles, as did Max Ernst's 'semi-automatic process' of *frottage*; the drip paintings of Jackson Pollock, like the work of Arp and Duchamp, employed gravity as a collaborator. The Fluxus group made chance an underlying principle of their world view, the accidental collision lifting us into something new, to effect the Epicurean swerve. Francis Bacon played roulette and painted like a gambler. Eva Hesse's plastic sculptures, because of the instability and perishability of their materials, had built into them the unpredictable shapes of their own decay.

Most things have a precursor. Warhol is no less of an artist because Duchamp got there first with industrial process and repeating image. Ernest Vincent Wright's novel *Gadsby* was published without the letter 'e' in 1939 (or almost: the book, a coy praising of youth's civic contributions, is a less rigorous exercise than Perec's and allowed in a handful). Marc Saporta's very pulpy *Numero 1*, published in 1963, is a novel without ordered pages.

In 1883 Lewis Carroll wrote,

For first you write a sentence,
    And then you chop it small;
Then mix the bits and sort them out
    Just as they chance to fall:
The order of the phrases makes
    No difference at all.

Thirty years later, Tristan Tzara published his instructions TO
MAKE A DADAIST POEM,

Take a newspaper.
Take some scissors.
Choose from this paper an article the length you want to
    make your poem.
Cut out the article.
Next carefully cut out each of the words that make up
    this article and put them all in a bag.
Shake gently.
Next take out each cutting one after the other.
Copy conscientiously in the order in which they left the bag.
The poem will resemble you.
And there you are – an infinitely original author of
    charming sensibility, even though unappreciated by
    the vulgar herd.

Carroll had been parodying the pretensions of modern
authors. Using exactly the same techniques, Tzara and his
Dada group, and such literary followers as Brion Gysin and
William Burroughs and David Bowie, were trying to get at
something more fundamental.

In the words of Hans Arp,

Chance opened up perceptions to me, immediate spiritual insights. Intuition led me to revere chance as the highest and deepest of laws, the law that rises from the fundament. An insignificant word might become a deadly thunderbolt. One little sound might destroy the Earth. One little sound might create a new universe.

Or, as Tristan Tzara wrote,

Dada is not at all modern. It is more in the nature of a return to an almost Buddhist religion of indifference.

Which leads to the Buddhist avant-garde composer John Cage, who used chance procedures to free him from the burden of intention.

'Yes, the whole thing is very determined,' Cage said about the processes that generate his compositions. 'But, those determinations were all arrived at by two different chance operations.' One of these operations involved – like a Leonardo prompt – the writing of musical notes suggested by the imperfections in the manuscript paper he was working on; the other was the casting of the *I Ching*.

(The precursor to this was a parlour game tradition. A booklet from around 1800 gave instructions on how to compose an infinite number of waltzes by means of two dice and a grid, 'Instruction pour composer autant de walzes que l'on veut par le moyen de deux Dez sans avoir la moindre conoissance de la Musique ou de la Composition': 'Instruction to compose as many waltzes as you want by means of two Dice without having any knowledge of Music or Composition'.)

Cage said, 'The highest purpose is to have no purpose at all. This puts one in accord with nature in her manner of operations.' He built a career out of actualising Duchamp's notion of 'the sounding sculpture', making something impermanent in space. His work seems to enact its promise of being without expressiveness or intent. And like Duchamp, with whom he played chess, Cage is very good at the sageish one-liner. Of his 'Freeman Etudes', he said, 'I think that this music, which is almost impossible, gives an instance of the practicality of the impossible.'

'I want to be free without being foolish.'

'I have nothing to say, and I am saying it.'

But then you come up against the lamentation and violence of 'In the Name of the Holocaust', for prepared piano from 1942, composed as a dance piece for his partner Merce Cunningham. The music starts quietly, almost respectfully, as if acknowledging its moment in history, becoming more expressive, louder and more fragmented as it goes on; solemn resonances clang against the silence that grows until the sound disappears into it, which seems appropriate, given the subject matter.

Except there is no subject matter. It is coincidence only that this piece was composed on the East Coast of the USA while the actions of the Final Solution were being perpetrated in Europe. The word 'holocaust', which means 'burnt offering', was not yet in general use to refer to Nazi atrocities. The title of Cage's 1942 piece comes from *Finnegans Wake*, its parody of the Lord's Prayer, 'In the name of the former and of the latter and of their holocaust. Allmen.'

The expressive quality of 'In the Name of the Holocaust', its 'meaning', which now can't be entirely extricated from it, has been bestowed upon the piece by chance. To put it another, more Duchampian, way, chance and history collaborated with their junior partner, the unwitting 'composer'.

Looking back at the compositional technique that he called 'limited aleatorism', Witold Lutosławski said that the origin moment of his work with randomising procedures came in 1960, 'a chance encounter with the music of John Cage through a radio broadcast… Those few moments were to change my life decisively…'

Cage was famous by 1960, and a progressive Polish composer would have been paying attention to the work of avant-garde Americans, especially if it was going to be played on the radio station in Warsaw. I have no doubt that Lutosławski listened to Cage's concert by intent, but by invoking chance here, the accidental, value-neutral moment, removed from anything that went before, the event somehow becomes 'good'. The chance event becomes a kind of blessing, a sign, a connection, a correspondence.

'Dada wished to destroy the reasonable frauds of men and recover the natural, unreasonable order,' Gabrielle Buffet-Picabia wrote.

This all comes very close to mysticism, the same spirit that led Jung to the *I Ching*, the revealing of the unity of all things, the occult patterns that hide beneath everyday phenomena. Stubbornly, the idea persists – even in the work and thinking of those who would seem to resist it most – that there is a secret pattern.

Even Duchamp made deliberate alterations and imposed some structures on to his supposedly chance operations. The notes given to Marcel in 'Erratum Musical' are lower than the other two parts, suggesting that some adaptation was performed to suit his lower voice.

And then there's his repetition of threes. The philosopher Richard Rorty in a spirit of awed exasperation referred to C.S. Peirce as 'one more whacked-out triadomaniac', in love with systems of threes. Duchamp's 'playful physics' is the work of another whacked-out triadomaniac. '3 stoppages' is made up of three measures generated by three separate drops.

'Erratum Musical' has three voices singing three melodies, three times. 'For me,' Duchamp told Schwarz, 'three is a magical number.'

There's an occult idea here, one thing corresponds to another, there are secret structures, the Bible text is code to be deciphered. Duchamp was interested in the Kabbala, in alchemy, in puns, in the seemingly accidental connections between things to reveal hidden truths. Is it only coincidence, for example, that there are twenty-five notes in Duchamp's 'Erratum Musical' and the same number of unbound sections in Johnson's *The Unfortunates*?

Dostoevsky was playing roulette in German casinos in August 1865, when *Daniel Deronda* opens. Theodor Reik, the pupil of Freud whose master confided in him his dislike of Dostoevsky, became Frank Ramsey's psychoanalyst in Vienna in 1924, at about the time that Duchamp was designing his bonds to finance his own roulette playing. Is there a secret pattern here? The Buffonist stylist Wittgenstein was so dissatisfied with a staircase that he and Engelmann had built for his sister Gretl's house that he entered a lottery in the hope of winning sufficient funds to rebuild it. Is that part of a secret pattern?

Among Frank Ramsey's many contributions to future thought is what has become known, long after the prodigy's early death, as the 'Ramsey theory'. This argues, from mathematical grounds, that complete disorder is impossible because, in his biographer's words, 'every large set contains a substructure with a regular pattern or cliques with a discernible pattern'. True randomness is impossible, there can only be unpredictability; and meanwhile there are always patterns forming out there in the chaos, if only we can see them, or make them.

*

In one of his interviews with David Sylvester, the painter Francis Bacon said,

I feel that now my luck has completely deserted me as a gambler, for the present time. Luck's a funny thing; it runs in long patches, and sometimes one runs into a long patch of very good luck. When I was never able to earn any money from my work, I was able sometimes in casinos to make money which altered my life for a time, and I was able to live on it and live in a way that I would never have been able to if I had been earning it. but now I seem to have run out of that patch. I remember when I lived once for a long time in Monte Carlo and I became very obsessed by the casino and I spent whole days there… I had very little money, and I did sometimes have very lucky wins. I used to think that I heard the croupier calling out the winning number at roulette before the ball had fallen into the socket, and I used to go from table to table. And I remember one afternoon I went in there, and I was playing on three different tables, and I heard these echoes. And I was playing rather small stakes, but at the end of that afternoon chance had been very much on my side and I ended up with about sixteen hundred pounds, which was a lot of money for me then. Well, I immediately took a villa, and I stocked it with drink and all the food that I could buy in, but this chance didn't last very long, because in about ten days' time I could hardly buy my fare back to London from Monte Carlo. But it was a marvellous ten days and I had an enormous number of friends.

DS – You were saying that for the moment you've exhausted your luck as a gambler. What about luck in your work?

FB – I think that accident, which I would call luck, is one of the most important and fertile aspects of it, because, if anything works for me, I feel it is nothing I have made myself, but something which chance has been able to give me.

Duchamp was the strategician who employed chance as collaborator, as procedure, and as gesture. Behind the theoretician of absence (of intention, of skill, of selfhood) was someone who knew how to play to the crowd, who autographed every action. Not long after his move to New York, he produced a *Wanted* poster advertising himself behind aliases, making himself more evident while pretending to disappear (R. Mutt, Rrose Sélavy). When he went to visit his sister in Herne Bay, shortly before he devised his randomly generated measures and sounds, he adopted the performance persona of a visiting Frenchman who played a lot of tennis and refused to speak English.

Bacon was ultra-retinal, a romantic vision of the Great Artist, making something out of the raw materials of nothing, and like Leonardo or Cage, using random imperfections of the surface to create shapes and colours that pleased him. There's a pleasing irony in comparing the haphazardness of Bacon's intentional approach with the sureness of Duchamp's unwilled chance procedures.

I wrote before that Duchamp was the artist of the twentieth century. But, he lost. Collectors and gallerists and the artists to come defeated Duchamp. Instead of destroying the boundaries between art and non-art, between artist and spectator, his work built up the myth of the Great Genius Artist even more. Anything Duchamp signed attracted value, not because of what it was, nor because anyone signing anything is an equally meaningless or meaningful act, but because *he*, Duchamp, had signed it, because he had selected it. A bicycle wheel or a urinal signed by Duchamp has the same cultural status as Joyce's *Finnegans Wake*.

Following Duchamp, to invoke a chance procedure was to gesture towards acts of genius. Francis Bacon went to Monte Carlo, because to spend time in Monte Carlo was a station on the way to becoming a great artist.

Looking back on his own Monte Carlo period, Duchamp wrote,

> You sometimes had to wait half an hour for the numbers to appear in succession of reds and blacks and the few weeks I spent in Monte Carlo were so boring that I soon gave up, happy to get out of it without loss.

Duchamp had tried to turn roulette into chess. He returned to chess, that game of perfect information.

*

Meanwhile, I was having a good run at poker. I was winning flips. Most of my 60:40s and 70:30s were holding up, and I was winning more than my fair share when I was on the wrong side of it. I was making final tables in tournaments at the Vic and I was running well in cash games at the Empire, where I would play in the afternoons after a shift of reading Duchamp and Machiavelli and Ella Lyman Cabot, whose work followed on from Charles Sanders Peirce's but who was much less of a triadomaniac and also the writer of much more transparent prose. After writing up my notes, I'd cycle to Leicester Square or Edgware Road to boost my Vegas fund. I was becoming a habitual winner again. Poker is an easy game when your nuts hold up and your draws come in.

And, superstitiously, I couldn't escape the feeling that the habits I had fallen into – only going to the cardroom if I was relatively content with the work that I had achieved that day, changing from my work shirt to my casino shirt, keeping to a routine of time, and emotion, even (it shames me a little to admit to this) eating exactly the same lunch every day and ignoring other people's supermarket lists that were piling up in a corner – were supporting my winning habit.

poker £2,670

CHAPTER 11

The Rice of Chance

Daring is the beginning of action, but chance is
responsible for the end.

Democritus

If we gamble, the heavenly gate will be nearer.

The Mahabharata

net total £1,384

Yesterday I bumped into my friend Scarlett on the Euston
Road in London. We're colleagues at the University of Kent,
and neighbours on the coast at Deal. But I was on my way
to the British Library and she was coming out of St Pancras
station to go to a meeting in Soho and our paths crossed. We
greeted one another, she reminded me that we were due to
meet for dinner at the weekend, and I told her my long-held
theory, that we only meet by chance those people whom we
were going to see anyway. Scarlett and I are accustomed to
see each other and therefore we notice one another, even in
a fresh context. Think of those friends we have lost, who
have fallen away, who pass by unknowingly, like atoms.
All the Mark Wilsons I have known. Joel Cooper. Nicki
Gordon.

Today I am back at the British Library, in my preferred position at the rear of Rare Books. Near me are youngish men in glasses whose relationship with their laptop keyboards seems to be one of attack. Opposite, a young woman wearing headphones is sketching a medieval rose with the aid of a magnifying glass. Beside me, an impeccably quiet man is reading about Nazi death camps. On my desk are five volumes of Petrarch's *Remedies*, Boethius's *Consolation of Philosophy*, *The Epicurus Reader*, Howard Patch's *The Goddess Fortuna in Medieval Literature*, Montaigne's *Essays*, Robin Waterfield's anthology of pre-Socratic philosophy, Casanova's *The Story of My Life*, Gerd Gigerenzer's *The Empire of Chance*, Gerda Reith's *The Age of Chance*, and a book on Chance and Early Modern literature, whose bibliography contains some of the usual suspects in the field, Thomas Kavanagh on chance as it applies to the literary text, Gerolamo Cardano on probability and the story of his life, Ian Hacking on the history of probability, Lorraine Daston on the Enlightenment and F.N. David on the intellectual history of chance. Also on my desk is a set of headphones to block out the force of the young men's attacks and the conversation bubbling between two excited elderly researchers inspecting microfilm at the monitors.

I make this loop through notions of fortune, histories of luck – today taking notes on Democritus and Aristotle and Epicurus and Lucretius and Boethius and Montaigne, and wondering how I dare to add my voice to theirs. Can I claim their insights for my own? I could be like one of those literary journalists who pass off as their own the best bits in the book under review, while condescending only occasional praise on their actual author.

Montaigne in chapter 33 (or in some editions 34) of book 1 of his *Essays*, 'That Fortune is Oftentimes Observed to Act by the Rule of Reason', gives some examples of 'all sorts of faces' of Fortune, the way 'she seems to play upon us'.

In 1326, Isabel, Queen of England,

> having to sail from Zeeland into her own kingdom with an army in favour of her son against her husband, had been lost, had she come into the port she intended, being there laid wait for by the enemy; but fortune, against her will, threw her into another haven, where she landed in safety.

He gives another example,

> And that man of old, throwing a stone at a dog, hit and killed his mother-in-law, had he not reason to pronounce this verse?: *Fortune has more judgement than we.*

Dutifully, putting the bad taste of this remark to one side, I set to work unpicking the ideas here, and their sources. The 'verse' is from the Greek dramatist Menander, who also coined the gambling expression 'the die is cast' – quoted by Julius Caesar when he crossed the Rubicon with his army. The image Montaigne uses of fortune 'playing upon us' is an echo of sonnet 1 of Montaigne's lost friend Etienne La Boétie ('Hélas, comment de moy ma fortune se joue!' 'Alas, how my fortune plays with me!').

This chapter, and its summaries from Montaigne that lead back to the Greeks, had been conceived of as a kind of introduction to luck's lineages, which the online randomiser has placed, maybe awkwardly, near the end. Belatedly, here come more of the Greek and Roman sources that still determine our thinking on the subject, and their medieval interpreters and reinterpreters, and maybe, finally, we will get to the poker game between theoretical physicists in the Bavarian mountains.

Montaigne's notion of luck is derived from Aristotle's. In the *Physics*, Aristotle defines the chance event as one that 'takes place when a man, performing an action purposively, achieves a result which he might have, but did not in fact, seek purposively, as for example when a man digs his garden to plant vegetables and happens on buried treasure'.

Aristotle's views were formulated in opposition to earlier philosophers, Democritus and Leucippus, the 'atomists', who described a universe in which elemental particles of matter fall within an infinite void. Luck in Aristotle's world happens when the individual is active – it's the unpredicted event that occurs when you're doing something else.

Democritus had written against Tyche, which he seemed to regard as a subjective phenomenon,

Men have fashioned an image of Chance as an excuse for their own stupidity. For Chance rarely conflicts with Intelligence, and most things in life can be set in order by an intelligent sharpsightedness.

In atomist theory, we're free to trust to luck through disregard for or ignorance of the laws of nature, which are given by necessity. Our Intelligence might move some things our way. And that's where our freedom ends. The world and its processes are a machine that we are powerless to change, and freedom is an illusion, like the appearance of colour.

In response to Aristotle's criticisms of the fixed nature of Democritus's atomistic system, Epicurus came up with what he called the *parenklisis* and which we know as the 'swerve': the spontaneous deviation of atoms from their fixed paths to collide with one another and set up new directions and possibilities.

Epicurus's Roman disciple Lucretius described the process,

While the first bodies are being carried downwards by their own weight in a straight line through the void, at quite uncertain times and in uncertain places, they swerve a little from the course, just so much as you might call a change of motion. For if they were not apt to incline, all would fall downwards like raindrops through the profound void, no collision would take place and no blow would be caused amongst the first-beginnings: thus nature would never have produced anything.

The swerve is needed for something to arise out of nothing. There would be no change without the swerve, which breaks 'the bonds of fate, preventing one cause from following from another from infinity'.

This is the same conceptual move that the Epicureans' unknowing English fifteenth-century descendants, the gamblers, made when they dethroned God and allowed chance into a previously locked and determined system of necessity and called it 'luck'.

As Epicurus wrote,

The wise believe that certain events occur deterministically, that others are chance events, and that still others are in our own hands. They see also that necessity cannot be held morally responsible and that chance is an unpredictable thing... As for chance, the wise do not assume that it is a deity (as in popular belief)... nor do they regard it as an unpredictable cause of all events. It is their belief that good and evil are not the chance contributions of a deity, donated to mankind for the happy life, but rather that the initial circumstances for great good and evil are sometimes provided by chance. They think it preferable to have bad luck rationally than good luck irrationally. In other words, in human action it is better for a rational choice to be unsuccessful than an irrational choice to succeed through the agency of chance.

As they say in poker – and all the best poker players are good Epicureans – *decisions, not results*.

In the Epicurean view, the gods, whether they exist or not, are indifferent to us – there are no intelligences determining our fates, only the necessity of nature, and the chance collisions triggered by the swerve, and the choices we make.

The early atomists, Democritus and Leucippus, along with Heraclitus, who wrote so persuasively about the permanent conditions of flux and change, are, in an act of forgetting and belittling, categorised among the 'pre-Socratic philosophers' of the 'fifth century BC'. This designation is a double anachronism, a double teleology, that they should be defined by being earlier than the teacher who inspired the philosophers who argued against them, and then be backdated by the notional birth of the founder of a death cult five centuries after they were dead.

Epicurus founded his Athenian school and community, the Garden, during the Archonship of Anaxicrates, in Year 2 of the 118th Olympiad. This is how his contemporaries would have dated it. In Rome, this would be known as the Year of the Consulship of Caecus and Violens, or 447 *ab urbe condita* ('from the founding of the city'). I'm trying not to call it 306 BCE.

The atomists and the Sophists, who took their philosophising from town to town, were particularly unlucky in their representation by Socrates' pupil, Plato, and by Plato's pupil, Aristotle. It was Plato and Aristotle who 'won', so to speak, ancient philosophy. The Platonic tradition was rehabilitated in the medieval West following translations of Plato and Aristotle from Greek into Arabic and a couple of generations later into Latin. There was something congenial to the Christian church about Plato's and Aristotle's views on appearance and essence and truth. The Sophists became a euphemism for empty argument, truthless 'logic', fine words that buttered no parsnips. It was they who had developed the idea of kairos, both in rhetoric and philosophy, the timely moment, taking advantage of the opportune event.

Epicurean communities abounded in the classical world. The communitarians aimed to live virtuously and moderately, in the spirit of emotional sustainability, freed from the anxiety of death. Unlike groups founded on Platonic lines, women and slaves were welcome to join. These communities were written out of history by the first generations of Christians, who were appalled at the notion that the soul dies with the body, that the atoms it is composed of drift back into the void, that it is the responsibility of the individual to live well and prudently in the moment, in the company of others. 'Epicurean' became a synonym for empty pleasure-seeking.

I'm persuaded by Epicurus and Lucretius. So was Karl Marx, who wrote his doctoral dissertation on the swerve. But I'm beginning to feel like Casaubon again, compiling a catalogue of interesting things people have said about luck, and striking one-sided alliances with interesting people who have also investigated fortune and risk and chance.

I move my things around – notebook on left, books on right, transparent plastic British Library bag that contains cap and scarf. One of the narrow young men glares at me in response to the rustling I'm making. He's glad to be given this opportunity: the glaree glares at the glarer.

A previous reader's bookmark falls out of Gigerenzer's *The Empire of Chance*. It is a small rectangle of white paper that has the single word *Train* written on it, followed by a full stop. Is this meaningful? How is it meaningful? Surely if I am taking this subject seriously, the bets placed, the meals I've eaten because of other people's supermarket lists, I must make it meaningful. This could be the moment of kairos that I mustn't let go of. Even though I have my bike tethered outside, perhaps I should make my next journey by train. Maybe this will be the fortunate message that will save my life – because I choose the train, the lorry that would have killed me on my route home takes its left turn without looking or indicating and it rumbles on and no cyclist dies beneath its wheels.

I check my emails again, even though I've already been checking them more frequently than I should, but the amount of time I'm staring into space has become excessive, and also I realise that I am looking at my neighbours too much, either in disapproval or curiosity, which is at risk of coming over as a bit oppressive. The woman who has been sketching the rose puts aside her pencil to consult something online, and her limbs make strange angles at her laptop.

Foolishly, I've allowed the book to close while puzzling over the scrap of paper, so I'll never know which pages it was marking – and maybe if there is a meaning to this moment, it might have been provided by a line or a paragraph of text. I'd been enacting an inadvertent *sortes Virgilianae* and I've missed the point, mistaking the direction for the destination, the sign for the signified.

Casaubon's chief mistake as a scholar was not to pay attention to the Germans, whose researches had superseded his own. Perhaps over in Leipzig or Berlin or Heidelberg or Vienna there are systematisers of luck who have left me well

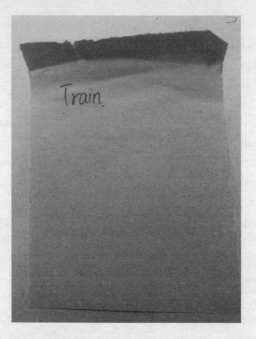

behind. All I can do in response, in the spirit of Nietzsche and Wittgenstein, is to go back to the words, trace where they have come from, see how they are used.

Kurt Latte points out that Tyche was a universal goddess, 'the bearer of the inexorable fate that is linked to all events'. Tyche is part of the structure of things, whereas Fortuna is particular, in the moment, changing from event to event.

Luck includes elements of both Fortuna and Tyche, but without her command. The processes have become desacralised. The mechanisms for consulting the oracle are now devices for gambling or talismans to console the superstitious.

The word 'dice' derives from Latin *data*, 'that which is given or decreed'. The rolling of dice is a discovery of that which has been given to pass.

There is dice-playing in the *Iliad* and the *Odyssey*. Plato attributes the creation of dice to Thoth, the god of writing and numbers. Sophocles claims they were invented by Palamedes to allay the boredom of his troops during the siege of Troy. Herodotus says they were invented by the Lydians to take their minds off a famine that lasted eighteen years.

They seem to have always been before us, with their ancient faces, the dots signifying numbers that come from a time before numerals had been invented.

I've been restricting these accounts to Western traditions, the lucky world I entered, its texts from the Greeks and Romans and Hebrews, but here we enter a universal prehistory. Dice dating back to 6000 BCE have been found in China and North and South America and Africa. *The Mahabharata*, the Sanskrit epic, has references to dice – Sakuni challenges Yudhisthira to a contest, 'the dice are my bows and arrows, the heart of the dice my string, the dicing rug my chariot!' Yudhisthira will stake a hundred thousand gold pieces, a thousand elephants, his slaves, his army and all his wealth in a single game. Finally he stakes the liberty of his brothers, his wife and even himself.

Dice have their own prehistory in *astragali*, the polished knuckle bones (actually heel bones) of cows and sheep, which the Greeks used for fortune-telling – astragalomancy – and to which the slang word *bones* for dice refers. These were replaced by *tesserae*, which we would recognise, six-sided cubes made of ivory, porcelain or marble, used in games and rituals.

The Roman game of *tabula*, a forerunner of backgammon, was played with three dice. (This is the reason why casino games are called 'table games'.) There were many other dice games played through the Roman Empire, such as the one in which players lost their wagers if they threw less than ten, and won if they scored more than ten. This is thought to be the game played after Jesus's crucifixion (John 19:23–4): 'When the soldiers had crucified Jesus they took his garments... and said... "let us not tear it, but cast lots to see whose it shall be".'

The warning against addiction to gambling in the *Mahabharata* is echoed in Tacitus's descriptions of the ruinous habits of German tribes: gambling with dice, recklessly risking their possessions, their freedoms, their lives, becoming literal slaves to gambling,

> They play at dice – surprisingly enough – when they are sober, making a serious business of it; and they are so reckless in their anxiety to win, however often they lose, that when everything else is gone they will stake their personal liberty on a last decisive throw. A loser willingly discharges his debt by becoming a slave: even though he may be the younger and stronger man, he allows himself to be bound and sold by the winner. Such is their stubborn persistence in a vicious practice – though they call it 'honour'. Slaves of this description are disposed of by way of trade, since even their owners want to escape the shame of such a victory.

During the Crusades, there were proscriptions on gambling according to rank and class. Anyone below the rank of knight was forbidden to play dice for money. Knights and clergymen were allowed to play, but they were not permitted to lose more than twenty shillings in a day, or else they were liable for a fine of a further one hundred shillings to the army's archbishops. Monarchs had absolute freedom to play, and no doubt to win, but their attendants were restricted to the same twenty-shilling limit, for which the penalty of exceeding it was to be whipped naked for three days.

The word 'hazard' derives from the Christian army's enforced halt in Syria in a castle called Hasart during the

Crusades. As in Sophocles' story of the ennui at Troy, this was a stay that was so long and boring the soldiers invented a game of dice, which they brought back home. Crusading soldiers were also said to have returned with playing cards, but these were more likely to have been imported from China to Europe by traders such as Marco Polo in the late fourteenth century.

Early decks of cards were crafted by hand on wood, copper and ivory as well as card and paper. Painters were commissioned to produce decks of cards, creating stencilled and woodblock designs that usually carried images of their patrons' noble families and lands. After the invention of printing, playing cards replaced dice as the most popular gambling tool.

Class hierarchies were expressed in the symbols and rankings of the four suits. In Italy, these were ranked as cups (the church), swords (nobility), money (merchants) and clubs (peasants). As hierarchies changed, swords swapped position with cups. Within each suit was the medieval army: king, knight (later replaced by queen), valet, foot soldiers. The joker was carried over from the tarot deck, symbolising, of course, the operations of a swerving unpredictable agency, uncontrollable, unexpected, which would soon be called luck.

Even without the joker, the gap between image and meaning allowed different, co-existing representations in the card deck. The king of spades might stand for King David, the king of hearts Charlemagne, the king of diamonds Julius Caesar, and the king of clubs Alexander the Great. But these identifications shifted and were never quite certain. The queen of hearts represented Judith, from the apocryphal book of the Bible, one of those heroines who use their charm to decapitate tyrants. Or else she was Isabeau of Bavaria, the spendthrift wife of Charles VI of France. The jack of clubs was sometimes Sir Lancelot, sometimes Judas Maccabeus, the Jewish insurrectionist.

Heraldic tradition took over the suits: in France, *piques* (spades) represented the knights, *cœurs* (hearts) the church, *tréfles* (clubs) the yeomanry, and *carreaux* (arrowheads or diamonds), the vassals. Face cards now showed idealisations of the European family: father, mother, eldest son.

Much of the information here on playing cards comes from an 1887 book by someone who styled herself Mrs John King van Rensselaer. Mrs John King van Rensselaer was also an expert on society and crochet and is one of many female authors who have been my guides. The principal exemplars of luck are men – Dostoevsky, Thomas Bastard, Casanova, Ashley Revell and so forth – with only Joan Ginther among them. But its principal writers are women: the historian Lorraine Daston, the statistician F.N. David; the sociologist Gerda Reith on the psychology of risk.

*

The epigraph to Casanova's memoirs is *Nequicquam sapit qui sibi non sapit* (a misquoting from the seventh of Cicero's *epistolae familiares*, which is itself a misquotation from Ennius's *Medea*). It translates as, 'He knows in vain who does not draw profit from what he knows' – or, as the song has it, 'If You're So Smart, How Come You Ain't Rich?'

There are more games of cards in his memoirs than there are accounts of seduction and love-making. Casanova gambles at basset, piquet, biribi, primero, whist, quinze and, above all, faro.

As the literary critic Thomas Kavanagh writes,

Casanova's frequent descriptions of the intense excitement of watching everything ride on a turning card reveal his fascination with gambling as another form of the treasured concentration of time into one all important moment of risk where success or failure depends only on chance... Casanova is first and foremost a champion of chance, of fortune, of his resolute determination to play and risk, win or lose. As seducer and gambler, he lives entirely within the present.

Contrary to what Kavanagh says, success or failure at cards did not depend entirely on chance. In the French court, where the ability to withstand huge losses in public with great sang-froid was more important than to win, Dangeau and Langlée had been guaranteed to clean up, by playing prudently, with large bankrolls, and with consideration to the odds. Even at the court, and more so outside of it, card cheats were looking for every edge they could find. Playing at Aix-en-Savoie in 1760, Casanova cautiously insisted on a new deck at every deal.

And, having lost all his money playing faro in Venice in 1763, he was approached by the Milanese Antonio Croce, 'a skilful corrector of Fortune's mistakes... Certain that this well-known cunning fellow had not spread a snare for me and assured that he had the secret of winning, I was not so scrupulous as to refuse him my assistance and the profit of one half the gain.' In several days they made 10,000 sequins (Venetian ducats) between them. *

Skilful correction... If we're thinking generously, like Kavanagh – who does not mention this incident – this could

---

* It is difficult to precisely estimate a contemporary value for historic currencies, but this would have been equivalent to a minimum of £50,000 today.

mean betting in accordance with the rules of probability, learning the lessons of the emerging science of *le calcul des hasards* to find an optimal strategy to pursue. But it probably means cheating.

In the second volume of his memoirs Casanova writes about when he was a young man doing his military service in Corfu, and had been wiped out at the faro table one night by a Major Maroli. Maroli had taken pity on him afterwards and offered him some 'sage maxims'. Perhaps these were a set of principles to gamble by. More likely, they were the mechanics of sleight of hand, the tricks of 'le Grec' as it was then called.

They became partners.

> I helped him when he dealt, and he rendered me the same office when I held the cards, which was often the case, because he was not generally liked. Maroli handled the cards in a way that made people afraid of him, whereas I did just the opposite; also, I was very lucky, and, in addition to that, my manner was easy and I smiled when I was losing and looked unhappy when I was winning.

Like Dangeau, Casanova made it fun to lose to him. He dressed the part, with 'large rings, expensive jewellery, fine clothes and carefully groomed wigs'; he always carried the same snuffbox to the table, with its erotic miniature hidden inside.

'The thing is to dazzle,' he wrote in *The Story of My Life*. Casanova was the son of actors, and he was playing a part here, this was his costume he put on when he was gambling, sending the message that he was an aristocrat rising above fortune's wheel.

But he needed a Maroli or a Croce to correct Fortune's mistakes. On his own, he became ruinously immersed in the action. 'Why did I gamble when I felt the losses so keenly?' he asked. 'I had neither prudence enough to leave off when fortune was adverse, nor sufficient control over myself when I had won.'

It was that desire to be in the moment, when time is concentrated in a turn of anticipation that seems to stretch out indefinitely, even if it is only that split second before the roulette ball finds its slot or the final card is revealed or as the beloved turns and is about to smile or frown.

Boethius wrote about this in *The Consolation of Philosophy*. This is the *nunc stans* ('the now that stands'), the eternal now,

Nunc fluens facit tempus,
nunc stans facit aeternitatum.
[The now that flows makes time,
The now that stands makes eternity.]

The *nunc stans* is similar to the Taoist concept of unaction, 'when nothing is done, when nothing is left undone'. As Wittgenstein noted, 'Only a person who lives not in time but in the present is happy.'

It is this state, the heightened perpetual now, that the gambler finds even if they didn't know they were looking for it. Like Casanova pursuing his fortune, like Aristotle's discoverer of buried treasure, it is found while looking for something that isn't there, in the gambler's case, the secret pattern that underlies everything.

*

I had been reading Gerda Reith on Boethius. She quotes, 'Chance does not exist at all', but then the idea goes further: neither can anything else,

> the world is, I judge, completely empty. For when God directs all things into an ordered pattern, what place at all can there be left for randomness?

I was stunned when I read this sentence. *The world is, I judge, completely empty.* I didn't know how to take it in. I sat suddenly back and then forward again at my British Library desk, alarming the man beside me, who was reading about Belsen and surreptitiously feeding himself nuts. There's an awful cold vision lying behind this remark that would chill anyone. I couldn't quite credit it: Boethius, the link between classical and medieval thought, descendant of Augustine, was saying something so stark that it went beyond nihilism. Reith had thrown it in, as an aside it seemed, maybe because this was too awful a vision to be able to share it directly.

It still didn't feel right. We're not meant to luxuriate in the world, because that keeps our attention away from God, but all the same isn't it filled with His creation? I needed a German scholar to explain this to me. I did some research – which these days means typing the words into Google – and the cold receded. It was a typo. The astonishing statement, of utter emptiness, was the addition of a rogue 'l' into Reith's discussion of Boethius's term *inanis vox*, the misspelling of 'world' for 'word'.

It was the *word*, chance, that was completely empty.

Boethius was denying the validity of the concept of chance, as an 'empty word', a term devoid of meaning, because the cardinal virtue of Faith eliminates the need for any other principle to explain what is.

He was allying himself with those, like Aristotle or Laplace or Einstein, who believed that all things are knowable, and ultimately predictable; we are merely ignorant of some of the causes; we don't have all the data. As Roger Bacon wrote in the thirteenth century, 'If the experiment of the magnet in relation to iron were not known to the world, it would seem a great miracle.' Or as Hume wrote, in accord with Boethius, 'chance, when strictly examined, is a mere negative word, and means not any real power which has anywhere a being in nature'.

I put on my headphones to shelter from the sound of nut-eating and turned to Denis Lejeune's *The Radical Use of Chance in 20th-Century Art*. Its opening chapter, 'The Rise of Chance in Modern Sciences', was the sort of historical survey I had read before, the Enlightenment project of quantifying the world, the grand measurers, these Protestant actuaries calculating a numerical value for such moral and psychological categories as risk and fear. On the right-hand page headings the chapter title was repeated, except there had been a misprint so each one read instead 'The Rice of Chance', which was an attractive image, implying an organic process, the paddy fields where chance grows, the bowls we feast from or whose empty depths we gaze hungrily into.

*

After the theoretical physicists of the early twentieth century found their way back to an atomist model of the universe – that vision of elementary particles in a void – Niels Bohr disagreed with Einstein that there were some hidden variables that would account for the seemingly random trajectories of the particles discovered by quantum mechanics.

According to the theories of quantum mechanics, energy radiates in tiny, discrete units, rather than smoothly and continuously as in the models of classical physics. Bohr demonstrated that electrons jump from one energy state to another. Werner Heisenberg's Uncertainty Principle proves the impossibility of ascribing precise values to both position and momentum simultaneously – which replaces Newtonian certainty (and Enlightenment measuring) with indeterminacy and, therefore, probability.

In January 1933, Bohr went into the Bavarian mountains in the company of his son, Christian, and three of his colleagues, including Heisenberg. They took a hut in Bayrischzell on the southern slope of the Großer Traithen. As Heisenberg recalled it,

We lay on the roof, from which we had cleared the snow, enjoying the sun and discussing recent developments in atomic physics.

That night we played poker… Our poker game was of a rather original variety. The hands on which we staked our money were shouted out and praised to the skies in an attempt to outbluff each other. Niels saw this as a fresh opportunity for philosophising about the meaning of language.

'It is quite obvious,' he said, 'that in this game we are using language quite differently than we do in science. To begin with, we try to hide rather than bring out the real facts. Bluffing is part of the game. But how do we hide the real facts? Language may convey pictures to others that help to oust ideas reached by sober reflection and so give rise to mistaken actions. But what factors decide whether or not these pictures impinge on others with sufficient intensity? Surely not the loudness of our voices. This would be much too primitive a view. Nor is it the kind of routine persuasion a good salesman might use. For none of us are familiar with this routine, and we can hardly imagine that any of us would be taken in it by it either.'

Bohr went on to suggest,

Perhaps our ability to convince others depends on the intensity with which we can persuade others of the force of our own imagination.

This theory was proved in practice when Bohr drove out all the other players with the power of his betting.

When it was all over, and he proudly showed us his fifth card, it turned out it was not of the same suit, as he himself had wrongly thought. He had mistaken the ten of hearts for a ten of diamonds.

His bluff had carried such certainty because he hadn't realised that he was bluffing. A few days later, they played again. (They waited a few days?! They did have quite a lot to discuss, atomic theory, Democritus, cloud chambers, energy, theories of language.) This time Bohr suggested they play poker without cards.

The attempt was made, but did not lead to a successful game, whereupon Niels said, 'My suggestion was probably based on an overestimate of the importance of language; language is forced to rely on some link with reality. In real poker one plays with real cards. In that case, we can use language to "improve" the real hand with as much optimism and conviction as we can summon up. But if we start with no reality at all, then it becomes impossible to make credible suggestions.'

It is in the mistakes that are made, typos introduced in the printing process, mutations in the genetic coding, an excited physicist in the mountains misreading hearts for diamonds, where we see the actions of the swerve most clearly, where we find our escape from necessity.

And still we persist in this tendency to find patterns, to get mystical. The Ramseyan structures assert themselves from below, the writer tries to establish some order from above. Patterns occur, repeat, break off – and I find myself imposing some more patterns as we move towards Vegas.

The Money has been calling. The Professor is joining us on our trip. I should call him back. I've read too much today. I need to clear my head.

I show my bag to the guard at the exit of Rare Books to prove that I'm not stealing anything. I drink water at the fountain. In the interest of novelty, the smallest of swerves, I use the ground-floor toilet rather than the one on the first floor. And then, continuing the swerve, I go into the British Library bookshop and notice that they've started an imprint, republishing obscure twentieth-century English science fiction books, and then I take a further turn around the museum shop wondering if there's anything here for family presents. My head is still clouded with thought and loose trails of ideas and the absence of connections between things, and I should go further into the Sophists but maybe I'll save them for another day. I'm ambling (maybe shambling) past the queue for the till when a man turns and says, 'Dave'.

I was only known as Dave for a few years in my life, roughly when I was an undergraduate at university, which aids me when the man points to himself and says, 'Ed'. And his middle-aged features morph into those of Edwin Jones, who I knew when I was a student, and probably last saw in Brighton in 1984. He works for the NHS in Cardiff and has been in London for a conference, and has bought some things from the Library shop. I always liked Edwin Jones. Everyone did, but not because there was anything insipid about him (as Saint-Simon might say). He dressed well and there was a gentleness to him and he liked to dance and there was a clarity about the way he thought and spoke. Unlike most of us, he seemed to know who he was and was comfortable in his skin. He's older than he used to be, obviously, with close-cropped grey hair and a handsome, slightly battered face.

He tells me that sometimes he gets a feeling that he is going to see someone he hasn't seen for a long time, and he got the feeling yesterday and has been on the alert for an encounter like this one, but didn't know who it was going to be with, or from which period of his life. And I tell him that he has disproved my theory that we only bump into the people that we see anyway (as Joel Cooper and Nicki Gordon and all the Mark Wilsons go past unnoticed).

Edwin Jones has been vigilant for this. He asks to see my little finger and he shows me his, and he tells me he's glad that I haven't had it repaired, and I remember, more slowly than this reunion should require, that there were three of us who had had accidents to little fingers that had bonded us in some way.

We go on to talk of our children, and of our lives. My head is still cloudy from Boethius and Bohr and I'm slightly embarrassed to be wearing unstylish clothes that I had allowed to select themselves for me in the morning, shades of brown and blue that don't quite match. Ed is wearing a suit and tie, probably because that was the dress code for his conference, and he's something high up in NHS Wales, and I'm glad he's doing good works although he laughs off that suggestion. He had been a Welsh nationalist and a Welsh language enthusiast, and I suppose he still is. There's the same aura about him that I remember, a very attractive combination of sincerity and humour. He's one of the few of us who back then was who he pretended to be.

Before we part – an exchange of details, of email addresses – he asks me if I have written some things about Jewish identity, and I say, well a bit, but not really for a long time. Maybe he's thinking of the piece I wrote about twenty-five years ago, on the subject of my son's circumcision.

Going back up the stairs to Rare Books, I wonder if this means something. What would Montaigne do? I should make this book more about Jewish identity? It's already more about Jewish identity than I had expected it to be.

How can this moment change my life? I resolve to be more alert, to expect the swerve and, therefore, to see it, to be able to grab hold of Kairos's forelock before it has slipped past. And maybe the message – *Train.* – that is still on my desk when I return to Rare Books doesn't refer to a means of transport. It is a verb, not a noun, a reminder to prepare, to stay in shape mentally and physically, to be ready for kairos, to expect that every moment should change my life.

poker £1,818

CHAPTER 12

Group Luck

Further, as there is Providence and Fate concerned
with nations and cities, and also concerned with each
individual, so there is also Fortune, which should next
be treated. The power of the gods which orders for the
good things which are not uniform, and which happen
contrary to expectation, is commonly called Fortune,
and it is for this reason that the goddess is especially
worshipped in public by cities; for every city consists
of elements which are not uniform.

Sallustius, *On the Gods and the World*

net total: £3,202

The imperialist Cecil Rhodes woke his friend Albert Grey one
morning before sunrise at Government House in Bulawayo,
to share with him the conclusion that his thoughts had led
him to in the night, 'Have you never realised that you might
have been born a Chinaman, or a Hottentot, or that most
degraded of men, a Mashona? But you are not, you are an
Englishman and have subsequently drawn the greatest prize
in the lottery of life.'

By seeing all Mashona as belonging to the same degraded
group, Rhodes was able to justify acting against them, to take,
for the English crown and his own profit, their diamonds and

gold and to occupy their lands, and, if he felt it necessary in the service of this, to kill them. Actually, 'Mashona' was a designation that included at least five different tribes. And there would have been many Englishmen in 1887, in the East End of London say, where my maternal grandparents would shortly land, who did not feel that they had drawn any kind of prize in the lottery of life.

This chapter had been intended to be much longer than it is. I was going to find lucky and unlucky groups. I had imagined it turning on ironies about 'the luck of the Irish' and 'the Chosen People', those referents of hope for the regular recipients of cataclysm. But any notion of group luck rests on those same presumptions of unities and homogeneity that kept Cecil Rhodes awake at night: that any 'Englishman' is as blessed as any 'Mashona' is cursed. Like QPR fans who complain, *Why does it always happen to us…?* (The answer is that it doesn't; it only feels that way.)

The philosopher Nicholas Rescher begins his discussion of luck with the flight of the B-29 *Bockscar* on 9 August 1945. Three days before, *Enola Gay* had dropped 'Little Boy' on Hiroshima. *Bockscar* was heading for the armaments city of Kokura on the northern tip of Kyushu island, with 'Fat Man' in one of its bomb bays. *Bockscar* made three passes over Kokura but because of 70 per cent cloud cover and diminishing fuel, its crew followed the contingency plan of dropping the atom bomb on the port city of Nagasaki instead. Approximately 74,000 people died (a figure which doesn't include those who would later die from such cancers as leukaemia, which post-bomb Nagasaki had a four or five times greater incidence of than the general Japanese population). 'And what was an incredible piece of good luck for the inhabitants of Kokura turned equally bad for those of Nagasaki,' Rescher concludes.

Not quite 'equally'. The population of Nagasaki at the moment of explosion had been about 240,000. If we restrict the number of casualties to 74,000, that meant its inhabitants had roughly a 1:3 chance of not surviving the blast. It has been estimated that if the bomb had hit Kokura as planned, about 57,000 out of a population of 130,000 would have died, which is nearly 1:2.

The statistical view of life – 'disregarding the individual as such', Émile Durkheim wrote in his study of suicide – analyses groups, or 'populations', looking for movements that are measurable and, therefore, predictable within a range of probability. On a probabilistic measure in which 0 equates to certain death and 1 guaranteed survival, a citizen of Nagasaki on 9 August 1945 had a .70 chance of not dying from the effects of the blast before 31 December 1945. A citizen of Kokura's chances would have been lower: .56.

Even when it's a question of a bomb falling out of the sky on to your town, your chances of survival are not necessarily the same as your neighbour's. You happen to be out shopping, your neighbour is in the basement; their chances of survival are higher. Within groups there are different attitudes to risk. Men and women, for example, show statistically different behaviours in regard to seatbelt-wearing, bicycle helmets, speeding and drink driving, vaccinations and medical care. Perhaps your neighbour has taken greater precautions against this sort of thing. Perhaps she doesn't just happen to be in the basement; maybe she has taken to living there, while you're on the top floor, looking out at the spectacle, and you've never worn anti-radioactive protection or reinforced your walls. Statistical calculations try to take account of stratifications within populations, but the obsessively cautious twenty-one-year-old man, who always wears his seatbelt and never drink drives, still has to pay the same insurance premium as others in 'his' group. No group is homogenous.

Group luck only truly applies when there is no escape from within it, and when there are no distinctions between its members. *The luck of the Irish... The Chosen People...* those characterisations of groups that traditionally have suffered more misfortune than fortune. Leaving Ireland in famine or crossing the ghetto line, or climbing down from the train that has taken you from somewhere bad to somewhere unimaginably worse, and you're on your own. It becomes your luck again.

*

A few years before he died, I sat down with my father to ask him questions about the most perilous times of his life.

I was trying to find out the answers to three questions: What happened to Izio Flusfeder between the years 1939 and 1944? Was there something intrinsic to him that enabled him to survive these events? Did he become who he was as a result of those experiences?

The Joe Flusfeder I knew was a man ferocious in the pursuit of his own advantage. He had the most acute systematic intelligence of anyone I've known. His relationships to other people were predicated on their inevitably disappointing him in some way.

In the summer of 1940, a homesick eighteen-year-old from Warsaw who had had enough of the Soviet Union and wanted to get back home – with no news from the place that he was trying to get back to – and still wearing his Polish double-breasted suit, was one of about eighteen hundred people crammed into cattle carts headed for Siberia,

There was a big huge warehouse there. Shelves. Like four floors of shelves. And we were sleeping on those shelves for like two days. And in the meantime the Russian prisoners would be robbing us of everything we had.

From there they put us on trucks. Six people on board, sit with their legs stretched out. And the next six would sit on their lap. Then the next six on their lap.

They dropped us at the end of the railroad. It was in the middle of the forest. *You better make yourselves comfortable because you're going to be here for the rest of your lives.*

And we'd start a fire and sleep around the fire. And in the morning they'd give you a cup of watered-down soup with a pickled tomato floating in it and a slice of bread. And you were on that diet for weeks. Very dark bread, very wet bread.

The only thing that was there was stakes, indicating the centre of the railway. Open fires, temperature below zero. We spent the whole winter out there. Suddenly you'd realise your shoe was on fire because you'd slept too close to the fire.

About twenty-three hundred prisoners. At that time they started dying like flies. Cold, dysentery. *Who Doesn't Work Doesn't Eat* was the slogan.

On this point at least, Bolshevism agreed with Christianity. Lenin had adapted the phrase from Paul's epistle to Thessalonians. It was put rigorously into practice in the Gulag.

There were twenty, twenty-one people in a work gang, a brigadier in charge of each one, chosen by the group. They gave us handsaws and axes and we were supposed to be cutting down woods.

I realised very fast that the norm, the number…

He is looking for the word 'quota' but can't find it. We're sitting in his midtown Manhattan apartment with its glass walls showing the city below, the American success story and his son from London. English was my father's third or fourth language and it's letting him down now that his pampered, inquisitive son is pushing him to remember his life when he was eighteen, a scrawny teenager in a freezing prison camp fighting for his life.

... of lumber, of cut down trees, I don't remember, like six inches off the ground, you weren't allowed to cut it higher, with a handsaw, it isn't easy, you know, two-man saws, and after you'd knocked it down, you'd have to chop down the branches, burn them, then cut the trunk into one-metre pieces and stack them. And at the end of the day they would have an accountant walking by with a measure, and he would measure for every brigade how much of that lumber did you cut down. How every brigade would make a quota, they would get nine hundred grams of bread. And you would get soup in the morning and soup at night. If you did half, you only got half, five hundred grams, and one soup a day. And if you didn't make half the quota you'd get three hundred grams of bread and just the watered-down soup in the morning. And if you didn't make, I don't know how much, you didn't sleep with the rest of the guys. There was just a square area, tiles of wood around it, so it was in the open, they'd throw you in there by yourself.

His expression is sort of wistful – *Look at me! Remembering all this!* – and neither of us can quite believe that this was his life, or rather the life of a former self with the same surname. He's using the word 'quota' quite often now, as if to demonstrate that any fallibility was only temporary.

The thing is, there were some of these groups who said, 'We only want the very strong ones', because if you made over the quota you'd get what they called a *pierogok*, which was *pierogies*, that was a bonus... the problem was, all those big workers dropped dead very very fast. Most of them would die in the night, they just wouldn't wake up.

And the rest of us quickly realised that you can't possibly do that. So the thing we used to do was cut down a couple of trees, the rest of the effort was spent carrying lumber that was cut yesterday and moving it to the new area, claiming that it was the lumber you cut down today.

Then you had actually to build the railroad, bringing dirt up in wheelbarrows. Running the wheelbarrows up these two by sixes.

There were only six hundred of us that survived.

And this is said with some quiet pride... I ask him about depression and the possibility of suicide, and wouldn't there be some people who just refused to go on?

I only remember one fellow who lost his mind, who had to be taken away... Most, were able to handle it. Either you handle it or you die. There's nothing in between.

You become dirty, you don't wash yourself, you don't take care of yourself. And sometimes I would faint, pretend to faint, so you would stay out of work for the rest of the day.

There were very few suicides. Most of us were unshaven, dirty, lousy, suffering from bloody dysentery. You don't want to carry on but you don't want to finish it off either. You carry on.

I started to ask impossible questions. How can people cope with enormity? How did he make his accommodations with these events that demanded the worst and the best from him? How do things change?

And then what changes, is that you become a senior prisoner and you know a few shortcuts. There was obviously a black market. The bosses within the camps... They were your masters, you were their slaves, whatever they wanted you would do. Get you up two hours early, keep you at work. Make you work when the temperature was lower. They were being measured by productivity.

Look for jobs, the kitchen for instance. You were eating! There was food. Get a job cutting bread. Get a job for one of the bosses, watching his fire at night. That was an excellent job... You're only looking after your number one.

You steal. Bread. From people that gave you the opportunity to steal from them. They wanted to work and I didn't. What can they do to you? Send you to prison? Call the cops?

The only thing that was punitive was if you don't work. That was the only criteria. Nobody was running away. There was really nothing to punish you for.

You become quite expert at it... You don't respect anyone there. But there's nothing to steal after a while.

And even now, there's some reflex, maybe a moral reflex, determining how he's telling this, using the distancing 'you' to describe or allude to some of his 'worst' behaviour, the single 'I' to refer to himself at his physically lowest, and humblest, and an inclusive plural 'we' when they were all let go together, and he might be briefly part of a group again.

There was a first aid station, what they would call a hospital. And I got very very sick, and they put me into the hospital. It's funny, it's very vague now, as far as time goes... how it happened that I ended up in that hospital. And I would keep snapping at that thermometer to keep it up. And I was in that hospital, I don't remember how long, a couple of weeks. And as I was getting better there was a Georgian doctor, who was a prisoner, who didn't know the Latin alphabet, so I would help him, you know, distribute medication, so he made me an assistant. And I was one of the sick ones, so I was getting better food than I would have got in my camp. And I think that was just before they released us.

Another one, he was the most, I don't know how to describe him, wonderful wonderful person, he was the director of the Moscow Philharmonic and got twelve years I think, had a girlfriend who stole his diary, and he got twelve years and they made him the manager of the hospital wing. Those were some of the nice people.

I suffered from scurvy. The only thing they would give me was, they would take beans and put them in water and when they sprouted they would give that as medication for scurvy.

For dysentery they would give you ground charcoal.

So you went through two winters there?
One, and the beginning of the second one.

Do you think you would have survived the second one?
I doubt it. I doubt it very much.

After Germany broke the Nazi–Soviet Pact, the Polish nationals were released. Izio turned down the opportunity to join the Red Army, instead asked for a train ticket going south, but they didn't give him enough money to pay for his fare all the way down to Totskoye, where a Polish army was forming.

So I was actually travelling under a seat... the Russian thing is that they have a woman in every car, a conductor, so it's not easy to get aboard without being seen.

Do you remember what you were thinking at the time?
How to survive. That was about it. See what happens, see what happens...

And why am I telling this? Probably for the same constellation of reasons that drew me to sit down with my father to learn some more of his story. To try to grasp who he was, to begin to ask if the person he was, was as a result of his early survival, or if one of the reasons he survived was because he was who he was. Because, even more than Ashley Revell, life had become reduced to either/or: either he works or he dies; either he comes up with a way to survive or he dies; and whatever he does, no matter how resourceful he might be, and how physically resilient, he has to get lucky. The odds of surviving in his group of transportees were lower than if he had been in Nagasaki in 1945, or a 'Mashona' in Mashonaland in 1887.

I would qualify a lot of those experiences as major and there is no doubt that a seventeen-year-old kid out of Warsaw versus a seventy-one-year-old man in New York, you know, it's a transition that I don't think anyone can define or even imagine. It's virtually impossible to say what would have happened if there was no World War Two. I'm still in Warsaw or whatever, going through a normal profession, going through life, would I be the same as I am today? I wonder. I don't think so. But how much different I would have been is a guess.

Those possibilities flash through one's mind... when I was young there was thought of me going to England for a while, yet I refused to look towards the future as a kid. I never thought about what was going to happen in a couple of hours, let alone what I was going to do in two, three years ahead.

A grim sort of *nunc stans*, faced by those who are pushed to the very edge of themselves in the struggle to go on living. When faced with the ultimate precariousness, there's no point making plans.

Hope and fear don't come into this. Not only are they not measurable, they're not even there. Hope and fear are my father's family back in Warsaw, his father going along with things because maybe they might improve. Hope and fear are his brother becoming a ghetto policeman.

Meanwhile, Izio Flusfeder found his way down to the south of the USSR, where Polish refugees were gathering under General Anders to form a battalion of what would become part of the British 8th Army. He was able to join up, despite the army's reluctance to take on Jews.

And?
And? I was lucky.

He was lucky to have left Warsaw in November 1939. He was lucky to have survived a Siberian forced-labour camp. He was lucky to have found his way to Anders's army, and to have been allowed into it. And then he was lucky to get through dysentery and all the other diseases, and the Battle of Monte Cassino. After being demobbed in Cardiff in 1946, he made his way to London. He was lucky to find work in plastics factories, where his claims to have experience as a moulding engineer were believed.

After the war my father met his uncle again, his father's brother. Both of them had adopted the same new name, George, to help shape their identity in the aftermath. Jerzy-George had changed his surname too, and his religion, 'because it was all going to happen again'. Izio-George didn't change Flusfeder, because even he allowed this much hope to re-enter his life, that somewhere someone else from his family might have survived too and so long as he kept his name they would be able to find him.

When he met my mother, at a Polish–Jewish ex-servicemen's dance, his English was so limited and his accent so strong that she misheard his 'George' as 'Joe' and that became his name.

A year later, they were married and in 1951 they emigrated to America where 'the pursuit of happiness' was being promised them. My father's understanding of this promise was very different to my mother's. He saw it clearly, that the American notion of happiness, and its pursuit, is not about finding contentment or pleasure. It is about luck. As Alexis de Tocqueville wrote, 'The whole life of an American is passed like a game of chance.'

Happiness's root is *hap*, the Middle English word derived from the Scandinavian languages' word for good fortune. The Enlightenment legislators drafting the Declaration of Independence were offering the possibility for each American to have fewer impediments in the way of realising their own relationships to luck, to opportunity, to the chance to catch Kairos by the forelock.

In 1967, Trudy, who had continued to believe that the pursuit of happiness meant finding contentment, some kind of personal fulfilment, returned to London with their children. Joe remained in the US. He might sometimes offer a 'how' he had survived, and prospered, but never a 'why', other than 'dumb luck'... *the series of simple coincidences...*

I was in the infantry. At that point I developed what was referred to as chicken blindness. Makes you blind at night, so as soon as it gets dark, you would see the window but you wouldn't see the wall. This is a vitamin deficiency, probably [actually a fungal infection, ocular histoplasmosis]. And, I also at the same time had dysentery. And we came to the port of Pahlavi [now Bandar-e Anzali], I remember we were coming on the beach, and it was just open, palm leaves woven into a top and nothing on the sides, and the problem was I had to keep on 'running', it was bad enough during the day but at night I couldn't find my way. So I remember at night I eventually end up lying down next to the latrine. And then the following day, maybe that same day, they were giving us uniforms, they deloused us, and I got a dish of sour milk, milk I didn't have for a couple of years. And I got two dozen hard-boiled eggs. And I bought a handkerchief. That was the three things I bought. And one of those twenty-four eggs was rotten. But I finished twenty-three eggs at one sitting and the dish of sour milk. And that night I thought, I'm going to die anyway, so I might as well die with a full stomach. And early evening I fell asleep right near the latrine. And I stayed there all night. Woke up the following morning, no problem, no more dysentery. I still have chicken blindness, that takes me a while to get rid of.

The things he might have done, the edges of his character that he had to recognise, to be made aware, one hard-boiled egg after another, of the limits of himself, the things he might be capable of doing... (*You steal. Bread. From people that gave you the opportunity to steal from them.*) If you don't seize kairos, you die. Fifty, sixty years later, he could see it all so unsentimentally clear, without spite or sympathy, or belonging.

*

I once taught in a prison and could see the unlucky boy in each of the unlucky men who were sitting in front of me. A group of boys commit a crime, break into a house, steal something, break things, they run away, a policeman gives chase, and one (probably the slowest, or the most myopic, that's how the odds fall) trips over a single uneven paving stone, and as he gets up, a stranger inadvertently, fortuitously, stands in the way. He watches his friends running off, the golden one among them is already out of view, and maybe he, our one, is doing it out of some obligation to his friends, he may be someone's brother, who wasn't even meant to be there, but the workmen who were repairing the pavement were lazy that day, a bird squawks, the stranger tries to let him pass, but the two of them move to the same side and block each other again, a life is changed in a moment by an unknowing intercession, and he is caught. He was unlucky, and maybe the word makes more sense as an adjective than a noun.

Writing this book has taught me a suspicion of the singly determining noun, the word that claims to encapsulate the whole. When I was at work on Chapter 9, I was not reducible simply to being a 'fifty-six-year-old man'. My friends are not merely the Money and the Professor. Calling Cecil Rhodes an imperialist or Casanova an adventurer or Wittgenstein a philosopher is apt, but insufficient. The actions that we are taking are usually evident. Who we are might never be and is seldom singular.

This project has been an investigation of luck, but I'm beginning to wonder if Boethius is right, and 'luck' *is* an 'empty word', but what might not be empty are the words *lucky* and *luckily*: the concept exists more fully as an adjective or an adverb. There is no external power or agency and nor is it a quality inherent in the substance of the thing or event. It's what Epicurus called an *accident*, something that's part of an object, without being necessary to it.

That's why the philosophers struggle so hard when they try to define it. Some moral philosophers say that *luck* is the term for events that we cannot control, and other moral philosophers find the flaws and holes in their argument and its illustration, which is really only a metaphor. We can't control the sun rising in the east but this does not amaze us each morning (and maybe it should). You don't say how lucky it is for you that it has (and maybe you should).

Luck is the operations of chance taken personally. The unit of identity can be the group rather than the individual, until the lucky or unlucky moment separates everyone again. Those two catch the plague from within a group of people who were all exposed to the same infection. The rest were lucky enough to pass through it. She survives. He dies, unluckily.

And when does an event begin? Probabilists define events as 'independent' or 'dependent'. An independent event is one whose probability is not affected by any previous events. A hand of cards, for example, begins precisely when the dealer riffles the deck from the previous hand. But even that has its dependencies.

The night before leaving for Las Vegas I was at the Vic with French Eric. He was playing a tournament, I was playing cash. French Eric had driven us there. I was getting tired and I had a lot of travel ahead of me and when he came over to my table, having been knocked out of the tournament, I was happy to leave. I'd just been in the blinds, so I said I'd play my button and I picked up my cards, which were 52, off suit. Not quite the worst starting hand in poker (which is 72, off suit), but not far from it.

There was an early raise and the prudent action of course is to throw away my hand. But I played those cards because the weather was bad and we had driven rather than bicycled that night and I wanted my lift home, and because I was tired, and I played that hand because French Eric happened to come over to my table just at the moment when I was picking up my cards, and I wanted to test French Eric's theory that his presence brings me good fortune at the poker table, and because this was going to be my last hand of the night, and I had had a winning session and I felt expansive and a little frisky, and I played the hand because I was on the button so I had position to three-bet the raiser. He raised me straight back, and now my expansive feeling abruptly left me, but I felt honour bound, in a stupid gambling sort of way, to call.

When the cards came down ace-three-four, my opponent had flopped top set, three aces, and I'd flopped a straight, the wheel, ace-two-three-four-five, which, given the preflop betting action, it was impossible for me to have, so all our money went in, and my hand held and I'd just, luckily, somewhat shamefacedly, won my biggest pot of the night.

*

This book has changed many times, as I might have said before, in the course of its making. In one of its original conceptions it was going to belong to the species of ruminative traveller's memoir that I often enjoy reading. Richard Holmes's *Footsteps*, for example, is a seemingly effortless blend of fact and speculation and observation and memoir, the author travelling to Italy to follow Shelley, and to France in pursuit of Robert Louis Stevenson as well as his own younger self. Or Olivia Laing's *The Trip to Echo Spring*, in which she's on the road in America visiting sites associated with her subjects' lives and works and drinking. My trip to Baden-Baden had been intended to be one of many.

I had played roulette in the same casino as Dostoevsky. I had walked through the park where he had come to Turgenev's house to petition the richer author for funds, which both of them knew would never be paid back, and which knowledge fuelled the contempt that each felt for the other – along with anger on Dostoevsky's side and pity on Turgenev's. But I knew all that already. It's in the letters and novels and reminiscences. In Baden-Baden I lost faith in the significance of place, in its ability to add value to my studies. By not travelling, I was losing the moments of observation that are enjoyable to perceive and make, and which add colour to any account, but they aren't integral, they don't really matter (in the Epicurean terminology, they are accidents, not substance) – so I didn't travel to Dorchester to see the building that has replaced the madhouse where Thomas Bastard died, or to Monte Cassino, where my father had fired big guns in the Second World War, and where Wittgenstein had been interned in the First World War, or to Midsummer Common the next time the fair came to town, to imagine Wittgenstein at the fairground with his eyes shut listening to the course of his coins down their wooden run.

I didn't even go to see Ashley Revell in the end. On the telephone he had convinced me that there was nothing more he could tell. He had gambled and he had won, I wasn't going to understand anything more by watching him drink and look at a menu and consume food. I could make my investigations without going anywhere more exotic than the British Library, with the result that there are more descriptions of neighbours at my desk than the quality of the light in places my exemplars had been.

Any excuse though will take me to Las Vegas. I'm flying to LA, and the Money and the Professor and I are going to drive to Las Vegas, where I will play poker. We might even go looking for Joan Ginther.

And on the Friday the thirteenth that would fall on my third day there I would be flouting every gambler's superstition I can think of in my final reckoning with luck.

poker £957

CHAPTER 14

Friday the Thirteenth in Las Vegas

One is compelled to wager, it is not voluntary, you are
in the game.

Blaise Pascal, *Pensées*

Then Jesus said to them, My time [*kairos*] has not yet
come. But your time is always ready.

John 7:8

net total £4,159

I used to know someone who had grown up in a small town
on the North Island of New Zealand. The town was populated
by descendants of Scottish Protestants, who had established
a place of sober hard-working respectability. On Friday and
Saturday nights, the young people would go to a barn outside
the town limits, where there would be music and dancing
and some would get drunk and fight each other. None of this
spilled over back into the town: no one would say anything
about the bruises on the butcher boy's face; and if a couple
had found an intimacy during a dance that wouldn't alter the
formality of their relations during the rest of the week.

This is how Protestant societies work. The civic spaces
are designed for polite, hard-working respectability, and the

young people let off steam and the sinners do their sinning in self-contained places outside town limits. The USA is a very Protestant society, and Las Vegas is its barn.

Actually, it's two barns, a couple of miles away from each other. The original one, Fremont Street, Downtown, is a congenially ramshackle place. The one-time Glitter Gulch is five blocks of casinos and bars and souvenir shops covered by a canopy, strung with zipwire and blasted at night with music and air conditioning and lights (*The Fabulous Fremont Street Experience!*). A couple of miles to the south, connected by streets of bail bond stores, is the Strip. This is the gaudy place of fountains and pirate ships and hangover movies and the 'Welcome To Las Vegas' sign, where high-rise casino-resorts stretch out along Las Vegas Boulevard. It began with a 1930s gambling roadhouse, the Club Pair-O-Dice, was built up in the 1940s and 1950s with oil and Mafia money, and properly established itself after the Cuban revolution shut down America's earlier playground in 1959. The mountains behind, and the intolerable heat, remind any summer visitor who is rash enough to stray too far from air conditioning, that this is a place in the middle of the desert without any reason to be, except for cupidity, profit, pleasure, need and water. Because of its spring, shepherds would come here to feed their animals. The Spanish called it 'The Meadows' – Las Vegas.

I've driven in from Los Angeles in the company of two old friends. Taking it in turns at the wheel, we followed Interstate-15 through the Mojave Desert, shimmer of heat, truckstops and Joshua trees and the occasional sun-blasted town. Both of my companions are Londoners who have been living in LA for over twenty years. One has made it big in Hollywood as a writer and producer of network television shows. The other is a professor of the history of science at

California State University. The Money and I had planned this trip some months ago. The Professor joined us at short notice, leaving his wife and two children behind. The Professor's wife had been unresisting, even encouraging. Because this is America, it is understood that men need to get together, to drive through the desert, to drink cocktails and argue about politics in the Bellagio bar.

On the drive, after the squabble over what and whose music we should listen to had been resolved – the driver's most-played tracks on Spotify, we agreed – I'd been telling them about Joan Ginther.

'I draw on a lot of women writers for the book, but she's like the only female case study, the only heroine. Maybe I should build her up some more? We could go look for her. She might still be living on the Strip.'

At this late stage, as sometimes happens, I am having a crisis of nerve, rethinking the book entirely, its manner and procedures. Perhaps I could make the climax of the book into an In Search of Joan Ginther. Or the whole thing, rename it The Quest for Ginther.

'Here's what we know. I keep thinking of her as "Joan" but her name is pronounced "Jo-Ann". She's won over twenty million dollars in scratch cards on the Texas lottery. The odds of doing that are something like eighteen septillion to one. In 2011 she had a condominium on Paradise Road, opposite the Riviera. By now I suppose she'd be living in the suburbs. Maybe Henderson.'

The Professor, who is usually correct on ethical matters, disapproves of this line of enquiry.

'She should probably be left alone,' he says.

'Good luck to her,' the Money says.

'Yes, you're right,' I say.

The focus of this trip is to play poker and to gamble on

Friday the thirteenth in Las Vegas. To feel lucky, psychologists tell us, is the main part of being lucky. When I had looked up to the departure board at Heathrow, my heart had foolishly lifted to see that the plane to Los Angeles was boarding at gate B42. The number 42 has traditionally been my 'lucky number'. This had felt like a good omen, as if the universe was benevolently revealing something of its secret pattern. And this is exactly the kind of thing I'm meant to be expunging in myself.

*

Las Vegas is one of the fastest-growing cities in the USA. It's also one of the poorest. In the spring of 2021, only LA had a higher unemployment rate. The crime rate in Vegas is high, and getting higher. But the biggest police anti-crime initiative that I have ever seen there was a clampdown on pedlars selling bottles of drinking water without a licence. They're a common street sight, almost as common as the time-share salespeople and nightclub promoters and Mexican porn-slapping families flicking cards advertising erotic services on Las Vegas Boulevard South. The tourists traipsing along the Strip in the desert heat are grateful for the opportunity to pick up some water. But as the *Las Vegas Review-Journal* reported of one family group who had been warned off by a security guard outside Planet Hollywood, 'Dolores Smith, 20, acknowledges that the water she and her cousins are selling for $1 is unfair to licensed businesses that overcharge.' This is a very Vegas usage of the word 'unfair'.

Life in Las Vegas can be brutal but it is self-evident: it's all about money. If you've got a dollar in your pocket, you're entitled to gamble. Anyone can wander into the high-end casino-resorts, and people do, streams and streams of them, to spend their dollars on slot machines and liquor and nightclubs and adrenaline adventure, drinking luminous cocktails from giant glasses, girls in tiny skirts and high heels, boys trying to act like high rollers, the sex workers in the casino bars, with the looks they send out that manage to be both candid and modest, *You're a discerning and attractive gentleman. You and I maybe could...?* and the disabled people roll slowly through the aisles between slot machines in wheelchairs and mobility scooters – because, as the recession deepens, the proportion of disabled people in Vegas has noticeably risen: Mammon has found its Lourdes.

Vegas has lost some of its swagger over the past few years. Less is given away for free. There are fewer comps. Everything costs more. Profits in the casinos of Macau are now three times those of Las Vegas, which need to protect their income stream from the likes of Dolores Smith.

Nonetheless, I still love Vegas, its over-calculated gaudiness, its relentlessness, the desperate haven it's made for smokers and gamblers and pleasure-seekers. In other contexts, I might find it decadent rather than magnificent that a resort in the desert that was once a small oasis has more championship golf courses than anywhere else in the world. The water comes from the Hoover Dam and is diverted from towns in Southern California that will have to go thirsty.

And I love that you can play poker here all of the time, with many hundreds of games to choose from at any moment in the day or night. Every cash table, it seems, has at least one of the following: a cocky young man wearing enormous headphones, an implacable white-haired gentleman who only raises with the nuts, an Asian-American who's a lively and dangerous opponent, and a ferocious old lady with dyed red hair who bets aggressively and whose ancient hands are covered with heavy jewellery and raised veins.

It is late, it is my first night in Vegas, I've left the Money and the Professor to argue at the Bellagio bar and I'm sitting jet-lagged and sleepless at a $1–$2 game in the Golden Nugget, where I meet the Spirit of Prudence. Prudence is a young Liverpudlian with good manners and immaculate clothes. He is dressed in pristine white Adidas sports top and shorts, aviator sunglasses and Wynn cap. He plays optimal poker, waits for his spots, and exploits them for maximum value. He laughs easily but not excessively, understands everything, has the telephone number of every cardroom manager stored on his phone, and seems to have no opinions outside of poker. He recommends the $1–$2 game at the Venetian as being the best way of building up my bankroll. 'So this is your trade?' I ask him. 'What else is there?' he says.

*

I'd remembered that Vegas is cold – the air conditioning is pumped up so high that it freezes the core – I'd forgotten how LOUD it is. Music is piped into every public space. Loss to me sounds like muzak and feels like the chill of air conditioning on jet-lagged skin.

I've been rereading Balzac's *The Wild Ass's Skin*, which I'd picked at random from the shelves at home where I keep my books relating to luck. I take it down with me to the breakfast buffet at the Golden Nugget.

The Wild Ass's Skin is the story of one of those doomed ambitious young men, like Hermann the engineer in Pushkin's 'Queen of Spades', who inhabit the wilder fringes of nineteenth-century fiction and require the intervention of magical forces to make their fortune accord to their idea of what they deserve in the world. Raphael de Valentin enters a casino in the first chapter, where he loses his final gold coin at roulette and then goes off to drown himself in the Seine. Out of some sense of tact or fastidiousness he waits for nightfall to cloak his death, in the meantime occupying what he expects to be his final hour in an antiquarian emporium where all kinds of treasures from all cultures and all ages are stored. The sight of a painting of Christ by his namesake Raphael almost pulls him back towards the world, but instead the image of Christianity reminds him that death is probably the answer – and then the ancient antiquarian shows him the shagreen, the wild ass's skin of the title, which will grant any wish but shrink in the process, diminishing the possessor's life force each time.

I had used to skim Balzac's description of the bohemians' dinner, to get more quickly into the gambling scenes and the story of Raphael's precipitous rise. It could be a description of Vegas – of all the excess, which, by definition, is surfeit.

Men who were usually discreet were telling their secrets to gossips who were not listening... By now almost all the guests were whirling in that delightful limbo in which the light of the mind is extinguished and the body, released from the tyranny of its master, gives itself up to the delirious joys of freedom. A few snatches of song echoed out like the tinkling of a musical box forced to grind out its artificial and soulless tune. Silence and rowdiness had become strange bedfellows.

That feels like a fair description of the buffet room of the Golden Nugget. I breakfast instead at Starbucks where I learn that Americans no longer say *Can I get...?* It's changed to *I need. I need a blueberry muffin... I need a pastry... I need an Americano...*

I need a double espresso, which I drink at the bus stop, feeling guilty. Prudence wouldn't have bought a double espresso for $5. Prudence would have waited an hour and ordered it at the poker table and all it would have cost was a dollar tip for the cocktail waitress. But at least the Spirit of Prudence takes me to the Strip by bus – $8 for a 24-hour pass instead of $24 each way for a cab. I meet Ray, who used to be a musician, and a seller of Harley-Davidsons, and a poker player, who has been working as a dealer at Binion's for the past seven years.

Ray helps me out with the ticket machine and then he helps out an older couple with directions, and I praise him for being a good Samaritan.

'It's more a karma thing,' he says. 'It's like a casino floor. Everything's recorded.'

The Money and the Professor are planning to take a helicopter ride over Red Rock Canyon. I head to the Rio to play poker.

One of the beauties of poker is that anyone can play it, at any level. You don't get recreational players entering Wimbledon or the Open golf championship. But you do get them, thousands of them, at the World Series of Poker. All it takes is money.

In this case, my entry fee to a $1,000 side event was being paid by the Money, who might have an inflated opinion of my poker capacities. Staking arrangements are common in poker, with the player, as the phrase goes, selling off pieces of themselves.

After the first twenty minutes or so, I am, as they say, in the zone. I know exactly where I am in pots, I know which opponents I can bluff, and which will be unable to steer away from confrontations. It is clear who the good players at the table are, and, therefore, who I should try to avoid, and who

I should target. If their chips are up for grabs, I should be the one grabbing them. I feel at the top of my game, and my form, when I may prove, at least to myself, that I can function, even thrive, at any level and in any company.

As Al Alvarez reminds us, poker is 'social Darwinism in its purest, most brutal form: the weak go under and the fittest survive through calculation, insight, self-control, deception, plus an unwavering determination never to give a sucker an even break'. I am feeling so in control that I even have space in my heart to feel sorry for the gentleman at the other end of the table. He is thick-set with a kindly face and a white goatee that matches the colour of his cap. He is shaking, unmanned by nerves. Maybe he is a wealthy tourist who entered the tournament on a whim, but he has neither the stomach for it nor the skill. Any time he forces himself to play a hand, the agony of the event is written on his face and body. He gives his chips away, some to me, more to the clever taciturn Australian on my right, and when he has lost them all, when his tournament life is over, the relief of it returns him to some kind of version of himself.

There are over 4,500 entrants to this event, which will last for four days, with a first prize of $654,000. I'm not dreaming of this yet, or even really of surviving long enough to get past 90 per cent of the field and into the money. (Although it has crossed my mind that it would be gratifying to reward the Money with money.) At this stage the plan is to accumulate chips, with the thought of having enough to put me in some kind of decent position going into Day 2. I feel confident, I am on top of things.

Dostoevsky was right. If I maintain my discipline and emotional equilibrium, then I am a winning poker player.

London is a long way away. Kent is a long way away. Everywhere is a long way away.

Marcel Duchamp, when playing chess, 'would always take risks in order to play a beautiful game, rather than be cautious and brutal to win'. Here, I tell myself, is where I belong, playing a beautiful game.

And then my courage and my composure fail me and, therefore, in a Dostoevskian way, my luck. A new player arrives at our table, a glowering young man wearing enormous headphones and a baseball cap who sits down with towers of chips in front of him. I raise in middle position with pocket tens. He reraises in the dealer position. The flop comes down jack high. I check, he bets, I raise, and he reraises, putting me all in for the rest of my chips. I look at him, he glowers back at me. He's still organising his chips into their towers from all the racks he had needed to carry them over from his previous table.

I had put him on ace-king. Possibly, he has a big pocket pair, higher than my tens. He might have ace-jack. Or – and – he is playing position. The later you are to act in a betting round, the stronger your hand becomes. When you're the last to act, you have leverage. If you have mountains of chips, you have greater leverage. I suspect I am winning. I ask for time. My instincts tell me to call. I fold.

Poker is perhaps unique in that you are betting on an event that has already happened: the deck of cards has been shuffled and dealt; as more cards are revealed, more information is available. Playing a game of incomplete information, part of the agony is never finding out the answers to questions. I suspect that I had the better hand, but I will never know. Even if I were to ask him, drag him out from under his headphones, he would probably lie.

Poker is also a test of the processing power of the brain and the emotional discipline of the player in response to new information and fresh stimuli. I am still beating myself up over the previous hand when I overplay the subsequent one, committing all my chips in a to-ing and fro-ing of action with the Aussie; and when it is over, I have top two pairs, he has a set of jacks and I am out of the tournament. I had felt where I was, I had *known* where I was, but I was still off-balance from the earlier skirmish, and committed a kind of suicide. It only takes a moment – two hands, five minutes – to switch from being on top of things to taking the shameful walk away to the exit door. It happens all the time. I didn't like that it was happening to me.

I console myself with a cab downtown and go into the Golden Nugget and, remembering Prudence, in a half-baked adrenaline way, instead of ordering a whiskey at the bar, I sit down at a table with a short-stacked $100-worth of chips. The whiskey is slow in arriving and, genially splashing around chips, I am about $30 or $40 down by the time it does come. Soon I have got just $30 of my original buy-in. The table is fun, there are characters there, we're talking about luck and tomorrow, Friday the thirteenth, and the memory of my tournament shame recedes and I buy in for another couple of hundred dollars and order another whiskey.

There is a live-wire on my right. Originally a New Yorker, he is used to a much bigger game than this, and he knows everyone and seems to know just about everything. On my left is a large Glaswegian who tells me he is a drug dealer. I put the questions to the table. Does anyone have any particular relationship to luck? Across is a drag racer, whose only superstition is that he gets into his car for a race from the right-hand side. If he happens to get in from the left, he gets out and goes in again from the other side.

What do they think about Friday the thirteenth?

The Glaswegian says, 'It's the day after the twelfth.'

The live-wire says, 'If people find something unlucky, then

I find it lucky. If people are scared of a two-dollar bill, then it's good for me. Succinct enough for you?'

Yes. Thank you. I colour up my chips and at the cage I ask to cash out in fifty-dollar bills. This surprises the cashier, slightly alarms her. She has to unlock a special drawer to retrieve them, because no one in Vegas wants fifty-dollar bills. Everyone knows they're unlucky.

*

So this is the set-piece day. I'm wearing the 'unlucky' colour green. I've bought the bag of nuts I'm going to eat at the table. Already, I feel free from sympathetic magic; the traces, left over from my mother's example, of superstition are gone. I'd left my hotel room key on a side table during the night, without a tremor.

At the Venetian, the $1–$2 game that Prudence told me about, I sit with a stack of fifty-dollar bills on the table beneath my chips, which means they're in play. I've broken one of the bills for a stack of $2 bills. When I tip a cocktail waitress with the first of those, the player to my left flinches. This isn't what it's for, I try to tell him, this is my luck, not yours, this isn't against *you*, but he doesn't understand: it's maybe my accent or the nuts I'm eating while talking, the fact that everything I'm wearing, apart from my shoes, is green, or just that the concept is too hard to grasp, that I might not be chiselling away for my own advantage over his, that this could be truly about my own relationship with luck.

Two to my right is an ex-military boy, a soldierly version of Prudence, who plays methodically and well, keeping records of every move and every game, playing position, making sure he raises preflop about 30 per cent of the time. The nervous player to my left is replaced by a genial type from Iowa, who likes to play small-ball, floating around in pots, see where and when he can pick them up. On my immediate right is another Englishman, who's on holiday with his wife, getting a cheaper room rate, the 'poker rate'. She wants them to see a show tonight.

Down the other end of the table is someone with head-phones and a lot of chips, who isn't quite as good a player as he thinks he is. The ex-military player is very dull to talk to, at least on any subject other than poker, but he understands how he plays, and he's working to improve. He knows who he is.

We're chatting down our end of the table. Neither I nor the Englishman nor the Iowan have ever gone to a show in Vegas. Why would you want to? the Englishman says, when there's *this*? The Iowan tells us he is playing more loosely than he usually does. He's on holiday; he assures us he's a much more dangerous player back home, and we believe him. The rest of the table doesn't like us so much. We're enjoying ourselves, and, in a small way, winning. And then comes my big Friday the Thirteenth hand.

The ex-military raises. I reraise with pocket kings. Down the other end of the table comes a big four-bet from young headphones. The soldier folds. Young headphones is a very tight, nitty player and I immediately put him on pocket aces. He'll call in position with ace-king. He might have queens or kings, but there are only two of those left in the deck, which makes it very unlikely. He either has aces or queens. I call. The flop comes down with a king and two hearts. I've got the king of hearts, so I've got top set and a back-door draw to the second nut flush.

Headphones bets, I raise, he calls. The turn is another heart. I bet, he raises, I go all in; he dwells up; finally he calls. I never do see his cards, but it must be aces, with one of them being the ace of hearts. He can put me on two hands: I have either AK (with the king of hearts) or KK. Either way, with a card to come, he's not dead, he can still outdraw me if he spikes the fourth heart on the river. The river card is a blank, I've doubled up in an $1,100 pot. I eat another nut. I resist the temptation to touch my $50 bills for luck.

I text the Money where I am, because I have enough now to return his stake. That isn't sufficient inducement to get him here. He and the Professor have gone to a shooting range today, to fire semi-automatic weapons in the desert. We arrange to meet for dinner.

I leave generous tips with multiple $2 bills and say good-bye to all my new friends. There's an Omaha tournament at Caesars, which is just a walk away. The eventual winner of the tournament is a human poker robot, who comes to the final table with a mass of chips and impressively hammers away and hammers away and has no personality whatsoever. I end up second, for a prize that is a little less than the entry fee to yesterday's WSOP tournament.

*

The Money takes us out late that night to a fancy steak joint at the Bellagio where, somewhat giddy with the food and the wine and Vegas, he orders us the most expensive grappa in the house to finish off the meal. The waiter mildly observes, 'That's a dangerous thing to ask for in a place like this.' When he brings us our drinks we ask him what's going to happen to all the food we'd left on our plates. The portions here are enormous, because one of the things we're paying for (or the Money is paying for) is surfeit.

We ask him if it gets distributed to the homeless and the waiter tells us that the establishment wouldn't dare do that, in case it gets sued by someone who claims to have got sick on the leftovers.

Our way out of the restaurant takes us past the entrance of the Bellagio cardroom.

'How much do you need to play a big game?' the Money asks me.

I look up at the board. There's a seat free in a $10–$25 game.

'How much would you need?'

'To sit down? You want to have at least a hundred times the big blind. But that would be playing short-stacked in a game like that. A minimum of five grand I'd say. Probably with another five in reserve in case things go wrong. But I've never played that high.'

'Well here's your chance.'

The Money takes out his wad. He licks his thumb and starts counting out hundred-dollar bills.

'It's late,' I tell him.

I remind him that I've drunk grappa, I've spent most of the day at a tournament, which isn't a good preparation for cash. It's hard to adjust. The emotional climate is different.

All these are true, these are all valid reasons not to sit down at a big cash game, to play at a level much higher than I've ever played before, but they're excuses, we all know they are.

Prudence shouldn't mean the same thing as timidity. And I am being over-cautious, finding reasons not to do the thing, not to take the opportunity that is being offered.

There are some other reasons that I don't say. I've already lost $1,000 of his money. There is a vanity at work here, I don't want to lose his good opinion of me, and of my skills in a world that he doesn't quite understand, which maybe makes him respect them all the more. And even though the Money likes splashing his money around I don't want to be further beholden to him. We'd once been in a clothes shop in Provincetown together, with our wives. I'd seen a shirt I liked but balked at the price. I don't pay $150 for shirts. (I will pay that without thinking as an entry fee for a tournament that I have little chance of winning but that's another matter.) He'd tried to buy the shirt for me, which I'd refused.

He pretends to understand, but he is disappointed in me. The Professor is disappointed in me. So am I.

It shames me to write this. Maybe if this conversation had happened the previous day or the following one my response would have been different. But this is the moment, here is now, and Kairos has slipped by and I hadn't even reached out to try to grab him by the hair.

*

I finish the night, in a pilgrimage sort of way, at Binion's, where the World Series began, in a cardroom which will soon close for ever. At first the game is so nitty it is unbeatable. I try to make the action but the action isn't there to be made. I'd hoped Ray the dealer would be here but he isn't. A saggy man from Cincinnati persists in telling me every detail of his itinerary from his trip to England several years earlier, as a dealer named Bruce, who looks like a long white fish, deals relentlessly slowly.

Red walls, TV screens with the sound turned off, photographs of past poker greats on the walls. This cardroom is dying and I feel as if I am dying too. There's an ex-cop from New York who takes an age to squeeze out his cards and who only calls me with the second nuts rather than raise. As Bruce's long white arms sweep across the table, I try to get a final conversation going about luck. A southerner tells of the time he went to start his PhD in computer science and in his first year, while he was away, his friends founded a start-up, which made several million dollars before he returned for his summer vacation. This prompts his neighbour, bearded, under a University of Las Vegas cap, to tell of the time his high school buddies set up a band which they invited him to join, but he too decided to go off to college and a couple of years later his sister asked him if the guys in Linkin Park didn't look kind of familiar. There's no rancour in any of this, no resentment at being unlucky, just a card player's sanguine reflection on the way things go down. As the conversation continues, people relax, loosen up. The man who could have been in Linkin Park absent-mindedly loses a pot that maybe he shouldn't have got involved with, and the chips come my way.

I take a break, go outside with a cup of coffee, try to see beyond the lights to where the mountains will be, and I wonder, if life had been different, might I have ended here.

The biggest lie we tell ourselves is that things are just the way they are, and that they stay the same. We pretend that this place will be here for ever, your love is constant, my heart is true. Parents die, children change, loves grow cold, servants turn against their masters, buildings rise, towers fall, loves renew. Hearts keep beating until they stop.

The human body is in a state of constant change. Its homeo-static mechanisms keep temperature, blood sugar, blood pressure, oxygen and so on at relatively constant levels. Every day we might die; every day the body successfully holds itself in balance and continues to fight off infections, until one day it doesn't.

Yes, luck is the operation of chance taken personally, but, crucially, it's an operation that takes place over time. That photograph of me is not true, because it takes the image out of time. The moment after it was taken is unknown, only to be guessed at, uninductively.

Superstition, Wittgenstein told us, is the belief in the causal nexus. I'm finally free of it. I'm standing imagining the mountains in this particular *nunc stans*. Because a driver didn't wear his safety harness or because Ulysses Grant was a bad president or because Bugsy Siegel was shot with three fifty-dollar bills in his pocket has no connection to this moment.

Even the prospect of the sun rising tomorrow morning does not have a probability of 1. It's more likely than most things, more likely, say, than a previously healthy fifty-six-year-old man dying within the next twenty-four hours, it's more likely than winning the lottery, but it can't be certain. Nothing is certain. There is always room for luck to be involved.

The credo, if I am going to offer one – and I probably should – is to *take the chance*, the unlikely or obvious connection, the sudden joining up of dots to make a pattern, whether it be fanciful or not, an image drawn of a future frame through which we will hurl ourselves, after we have remembered to stop saying no.

Luck does not exist as a force external to us, intervening between desire and its consummation, to get us the things that we want; but it's the word we use to describe change, the difference in state, from one place to the next, from one moment to the next, *eventuality*. Luck is the swerve, the kairotic moment, and it doesn't just happen to us – it's the moment that we might grab, sensing an opportunity, without knowing the future we are exerting into being.

The last time I saw my father, we knew it was our final meeting. He was dying. We had taken our leave of each other and he had one more thing to say, which was, *Good luck*.

At the time, I was disappointed. I had been looking for something grand, something equivalent to the secret words whispered to Moses on Mount Sinai, the revelation of a previously hidden truth. I'm finally beginning to realise that it was as grand a blessing as he could have given me.

And now here I sit, blood chugging gamely through smoke-damaged arteries. Cells are dying, infections invading the body, the organism is fighting wars on multiple fronts; viruses, bacteria, pre-cancerous cells flourish and die. And meanwhile I try to create a sequence of sameness and samenesses in the face of constant flux and change. Everything changes, perpetually – held into its notional place by calling it *me*, by 'my' name.

I'm not a high roller. I'm more timid than I would have ever chosen to be. As Marcel Duchamp wrote during his time in Monte Carlo, 'I'm neither ruined nor a millionaire and will never be one or the other.' I might be on Vesuvius, but not by the open mouth of the volcano. You can find me on the down slopes, taking money off the tourists, that's where I'll be.

poker £1,242
TOTAL £5,311

Acknowledgements

There are so many people who have helped with 'The Luck Book'. I wish I could properly thank my late agent, the prodigious David Miller. But I can thank my current agent, Matthew Hamilton; my patient and steadfast editor, Nicholas Pearson; project editor Katy Archer, and Nicola Webb, along with everyone else at 4th Estate; copyeditor Anne Obrien; Ros Porter, who originally commissioned it; friends and colleagues, who have been generous with knowledge and enquiry, some of whom might find themselves, lightly disguised or not, in these pages: Mathew Gibson, Bruno Heller, Kevin Lambert, Roddy Campbell, Sheila Wedgwood, Scarlett Thomas, Edwin Jones, David Spiegelhalter, Natalie Galustian, Rod Edmond, Amy Sackville, Bernhard Klein, Patricia Novillo-Corválan, David Herd, Sara Lyons, Ali Thorpe, Adrian Smith, Anthony Holden, Louisa Young, Jon Walker, Kate Summerscale, Vybarr Cregan-Reid, Lucy Warburton; students of mine, particularly of Creative Writing Non-Fiction at the University of Kent, where we've been learning together how to do this; the staff at the British Library.

More images, and a bibliography, can be found at davidflusfeder.com.

This book is dedicated to four people very close to me who died untimely during the course of its writing: Felix Schroer, Deborah Orr, Michael Swift and David Miller; and to my family – my wife, Susan, and our children, Julius and Grace – who are my great good luck.